MOLECULAR BIOLOGY INTELLIGENCE UNIT 25

The CDK-Activating Kinase (CAK)

Philipp Kaldis, Ph.D.
Regulation of Cell Growth Laboratory
Center for Cancer Research
National Cancer Institute
Frederick, Maryland, U.S.A.

LANDES BIOSCIENCE / EUREKAH.COM
GEORGETOWN, TEXAS
U.S.A.

KLUWER ACADEMIC / PLENUM PUBLISHERS
NEW YORK, NEW YORK
U.S.A.

The CDK-Activating Kinase (CAK)

Molecular Biology Intelligence Unit 25

Landes Bioscience / Eurekah.com
and
Kluwer Academic / Plenum Publishers

Printed in Great Britain by IBT Global, London

Kluwer Academic / Plenum Publishers, 233 Spring Street, New York, New York, U.S.A. 10013
http://www.wkap.nl/

Please address all inquiries to the Eurekah.com / Landes Bioscience:
Eurekah.com / Landes Bioscience, 810 South Church Street, Georgetown, Texas, U.S.A. 78626
Phone: 512/ 863 7762; FAX: 512/ 863 0081; www.Eurekah.com; www.landesbioscience.com.
Landes tracking number: 1-58706-056-6

The CDK-Activating Kinase (CAK) edited by Philipp Kaldis/CRC, 124 pp. 6 x 9/ Landes/ Kluwer dual imprint/ Landes series: Molecular Biology Intelligence Unit 25,
ISBN 0-306-47438-7

Library of Congress Cataloging-in-Publication Data

CIP applied for but not received at time of publication.

Molecular Biology Intelligence Unit

Recent Volumes in this Series

The Jak-Stat Pathway in Hematopoiesis and Disease
Edited by Alister C. Ward

Telomerases, Telomeres and Cancer
Edited by Guido Krupp and Reza Parwaresch

Reactivation of the Cell Cycle in Terminally Differentiated Cells
Edited by Marco Crescenzi

Ceramide Signaling
Edited by Anthony H. Futerman

Nuclear Envelope Dynamics in Embryos and Somatic Cells
Edited by Philippe Collas

The CDK-Activating Kinase (CAK)
Edited by Philipp Kaldis

Caspases—Their Role in Cell Death and Cell Survival
Edited by Marek Los and Henning Walczak

Quasispecies and RNA Virus Evolution: Principles and Consequences
Edited by Esteban Domingo, Christof Biebricher, Manfred Eigen and John Holland

Cell Cycle Checkpoints and Cancer
Edited by Mikhail V. Blagosklonny

RNA-Binding Antibiotics
Edited by Renée Schroeder and Mary G. Wallis

Lysosomal Pathways of Protein Degradation
Edited by J. Fred Dice

CONTENTS

EDITOR

Philipp Kaldis, Ph.D.
Regulation of Cell Growth Laboratory
Center for Cancer Research
National Cancer Institute
Frederick, Maryland, U.S.A.
Chapter 2

CONTRIBUTORS

Nick R. Brown
Department of Biochemistry
Oxford University
Oxford, England
Chapter 6

Aiyang Cheng
Department of Molecular Biophysics and Biochemistry
Yale University
New Haven, Connecticut, U.S.A.
Chapter 2

Jean-Marc Egly
Institut de Génétique et de Biologie Moléculaire et Cellulaire
Université Louis Pasteur
Illkirch Cedex, France
Chapter 7

Jane A. Endicott
Department of Biochemistry
Oxford University
Oxford, England
Chapter 6

Robert P. Fisher
Cell Biology Program
Memorial Sloan-Kettering Cancer Center
New York, New York, U.S.A.
Chapter 4

Damien Hermand
Haartman Institute
University of Helsinki
Helsinki, Finland
Chapter 3

Anne Keriel
Institut de Génétique et de Biologie Moléculaire et Cellulaire
Université Louis Pasteur
Illkirch Cedex, France
Chapter 7

Edward T. Kipreos
Department of Cellular Biology
University of Georgia
Athens, Georgia, U.S.A.
Chapter8

Stéphane Larochelle
Cell Biology Program
Memorial Sloan-Kettering Cancer Center
New York, New York, U.S.A.
Chapter 4

Ji Liu
Department of Cellular Biology
University of Georgia
Athens, Georgia, U.S.A.
Chapter 8

Tomi P. Mäkelä
Haartman Institute
University of Helsinki
Helsinki, Finland
Chapter 3

Randy Y.C. Poon
Department of Biochemistry
Hong Kong University of Science and Technology
Clear Water Bay, Hong Kong
Chapter 1

Karen E. Ross
Department of Cell Biology
Yale University
New Haven, Connecticut, U.S.A.
Chapter 2

Vasiliki Tsakraklides
Department of Cell Biology
Yale University
New Haven, Connecticut, U.S.A.
Chapter 2

Masaaki Umeda
Institute of Molecular and Cellular Biosciences
The University of Tokyo
Tokyo, Japan
Chapter 5

Thomas Westerling
Haartman Institute
University of Helsinki
Helsinki, Finland
Chapter 3

Edward Winter
Department of Biochemistry and Molecular Pharmacology
Thomas Jefferson University
Philadelphia, Pennsylvania, U.S.A.
Chapter 2

PREFACE

Cell growth and cell division belong to the most basic events in biology. After induction of mitosis by the maturation promoting factor in the early 1970's, it took more than a decade to identify the gene products involved: the cyclin-dependent kinases (CDKs) and their regulatory subunits, the cyclins. The activity of CDK/cyclin complexes forms the basic oscillator that promotes transition from one cell cycle phase to the next and therefore drives the cell cycle. Regulation of the CDK activity turned out to be rather complex and includes phosphorylation of the activating threonine by the CDK-activating kinase (CAK). This book aims to extract and summarize all information about CAK by pointing out commonly accepted facts and unresolved issues. It will take the reader from yeast to mammals and describe all areas that CAK is thought to be involved in. This book is designed to serve newcomers to the field as well as specialists; anybody interested in cell growth, signal transduction, and cancer. I hope it will stimulate discussion of the published data and of the unanswered questions to generate new research directions.

The Chapters of this book have been written by experts in the field. In Chapter 1, Randy Poon presents an overview of CAK from marine invertebrates to humans. In these systems, CAK function is carried out by CDK7/cyclin H, a kinase that is also a subunit of the general transcription factor IIH (TFIIH) and phosphorylates the C-terminal domain of the large subunit of RNA polymerase II. The role of the reversible activating phosphorylation in Cdc28p from *Saccharomyces cerevisiae* is described by Kaldis et al in Chapter 2. We review genetic and biochemical data on the budding yeast CDK-activating kinase, Cak1p, and the PP2C phosphatases, Ptc2p and Ptc3p, that remove the activating phosphate from Cdc28p. In Chapter 3, Westerling et al present the model system *Schizosaccharomyces pombe* with the two kinases Mcs6/Mcs2 and Csk1. Both Mcs6/Mcs2 (a CDK7/cyclin H homolog) and Csk1 (a Cak1p homolog) display CAK activity. Larochelle and Fisher discuss CDK7 in *Drosophila melanogaster* in Chapter 4. Drosophila is the only higher eukaryote where the in vivo function for CDK7 has been demonstrated. CAKs have also been identified in plants and in Chapter 5 Masaaki Umeda reviews the different plant CAKs, where some resemble CDK7 and others are more Cak1p-like. The structural aspect of CDK activation plays an important role in the understanding of CAKs. In Chapter 6, Brown and Endicott summarize how activation of CDKs by CAK affects their structure. TFIIH plays a major role in transcription and DNA repair. In Chapter 7, Keriel and Egly describe the involvement of CDK7 in transcription and DNA repair as a subunit of TFIIH. Liu and Kipreos review in Chapter 8 the evolution of CAKs. This is particularly interesting when we try to understand the similarities and differences of CAKs in various species.

With all this information about CAK in mind, there are still many aspects to be investigated. One of the central questions are the physiological roles of activating phosphorylation, especially since CAK activity and CDK phosphorylation seem to be constant during the cell cycle. Another major question is the existence of two CAK systems, appearing either separately or in combination in different species. For example, CDK7-like kinases, Cak1p-like kinases, or representatives of both enzymes can be found in mammals, budding yeast, and fission yeast, respectively. Given the conservation of other components of cell cycle pathways in diverse species, we will need to find reasons for the differences in CAK systems. In addition to these questions, there are also additional functions and substrates of CAKs that we are only beginning to discover. Investigating CAKs parallel in several species will continue to be a powerful approach for new findings.

The completion of this book depended on the contribution of the authors of the Chapters and I would like to thank them for their effort. In addition, I thank Michele Pagano for encouraging me to edit this book.

Philipp Kaldis
July, 2002

COLOR PLATES

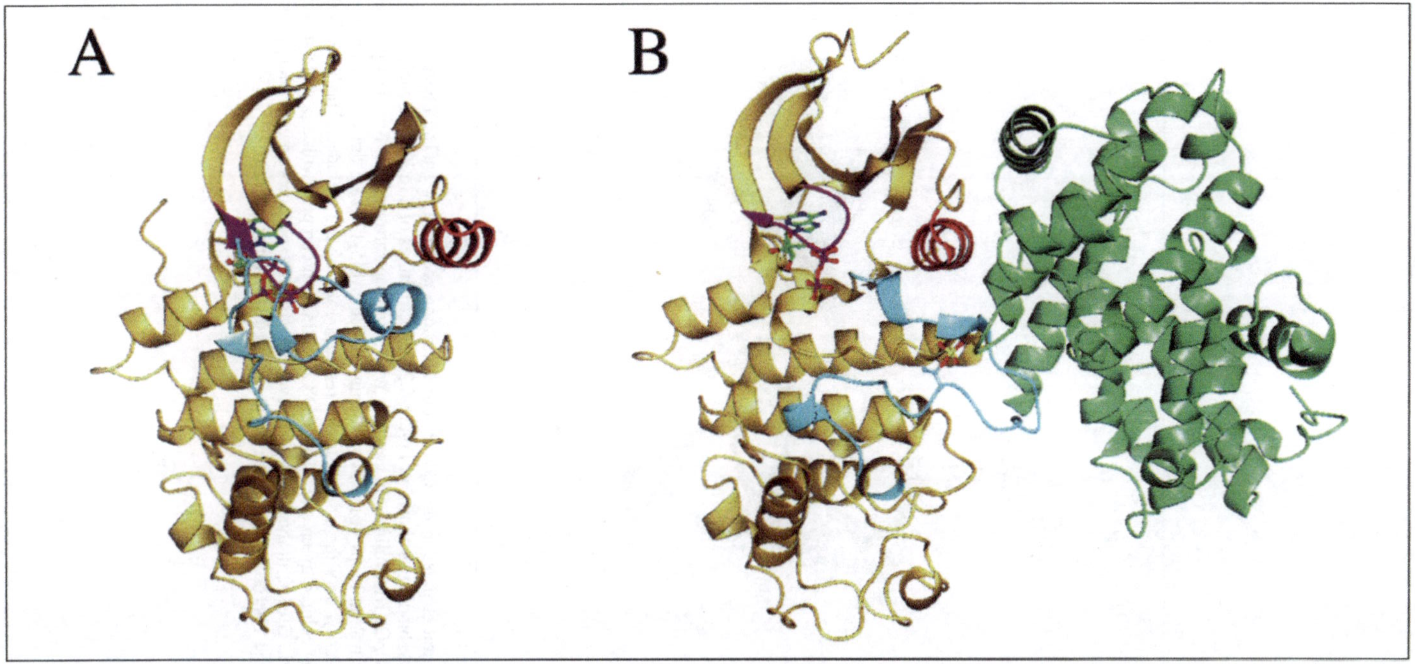

Figure C.1. CDK activation as shown in Figure 6.1. The structures of monomeric CDK2 (PDB code 1hck) (A) and T160pCDK2/cyclin A in complex with a peptide substrate (PDB code 1qmz) (B) were superposed using the CDK2 C-terminal lobe. The CDK2 and cyclin folds are drawn in ribbon representation and colored in gold and in green respectively. The CDK2 glycine loop (residues 11 to 16) is highlighted in pink, the C-helix (residues 45-55) in red and the activation look (residues 145-172) in cyan. AMP-PNP is drawn in ball and stick representation with carbon atoms in green, nitrogens in blue and oxygen atoms coloured red. The phospho-Thr160 side chain is shown. The figure highlights the large structural rearrangements within CDK2, particularly of the C-helix and of the activation loop that accompany cyclin binding and phosphorylation of Thr160.

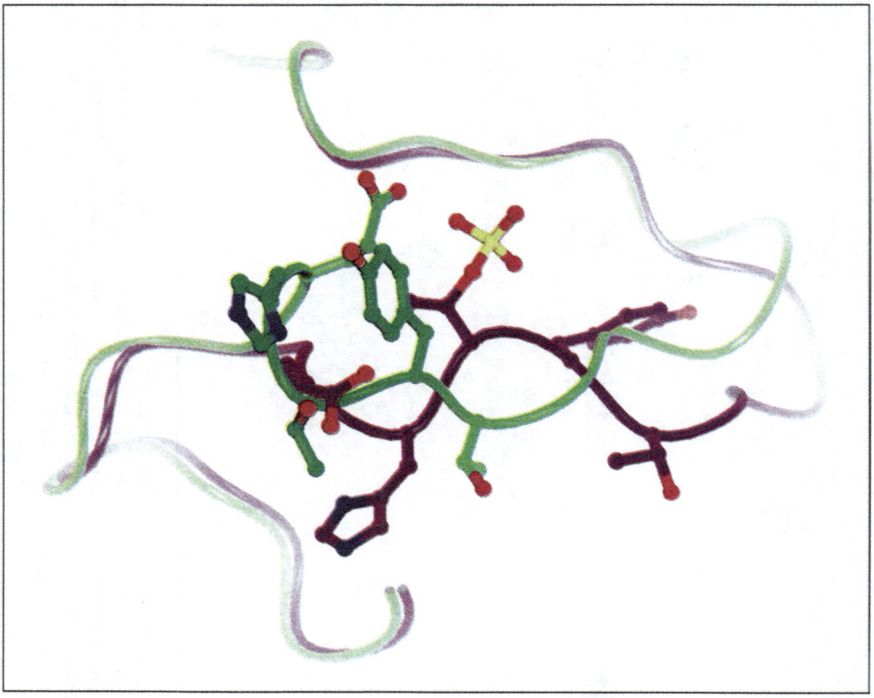

Figure C.2. Structural changes in the CDK2 activation loop accompany Thr160 phosphorylation and substrate peptide binding as shown in Figure 6.2. The structures of T160CDK2/cyclin A (PDB code 1fin) and T160pCDK2/cyclin A in complex with a peptide substrate (PDB code 1qmz) were superposed. The activation loops of the two proteins (residues 145-172) are drawn in worm representation and for clarity, only the side chains of residues 158-162 are included. The T160CDK2/cyclin A structure is colored green and T160pCDK2/cyclin A in complex with a peptide substrate is shown in magenta. Glu162 in the T160CDK2/cyclin A structure occupies an equivalent position to the phosphorylated Thr160 residue in the fully-activated structure (magenta).

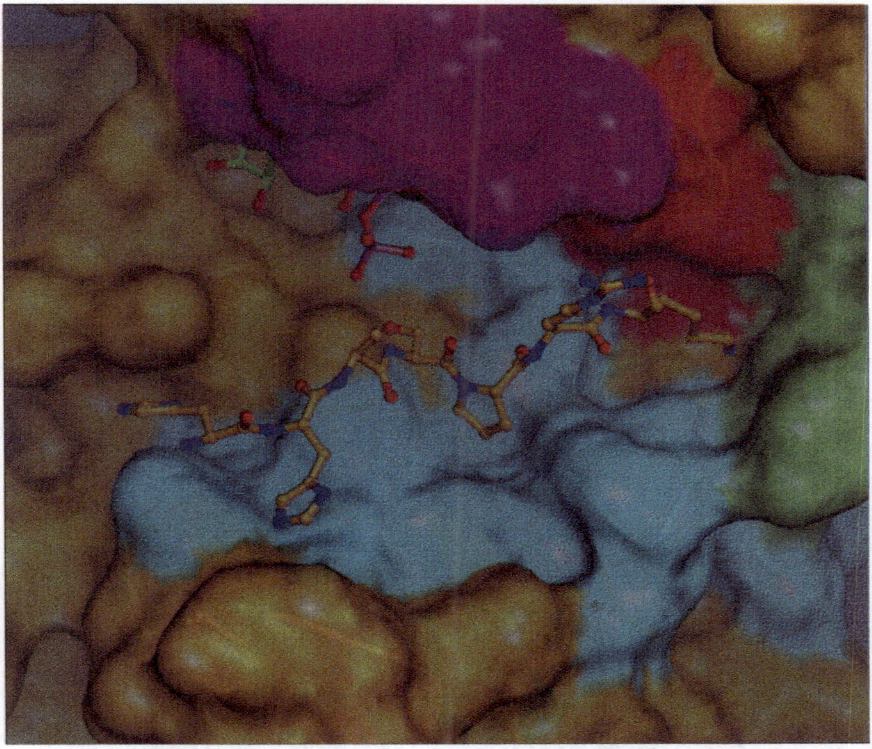

Figure C.3. The binding of the consensus substrate peptide to CDK2 as shown in Figure 6.3. The structure of a phosphorylated CDK2/cyclin A/AMPPNP/Mg^{2+}/peptide complex (PDB code 1qmz) in the vicinity of the catalytic site is shown. The CDK2/cyclin A molecular surface is colored according to the color scheme employed in Figure 6.1: CDK2 in gold, cyclin A in green, and residues from the glycine loop, C-helix and activation loop coloured pink, red and cyan respectively. ATP and the substrate peptide are drawn in ball and stick representation with the carbon atoms in the ATP molecule coloured green and those of the substrate peptide coloured yellow. The CDK2 binding pockets for the substrate amino acid side-chains of histidine (P-2), proline (P+1) and lysine (P+3) are clearly seen.

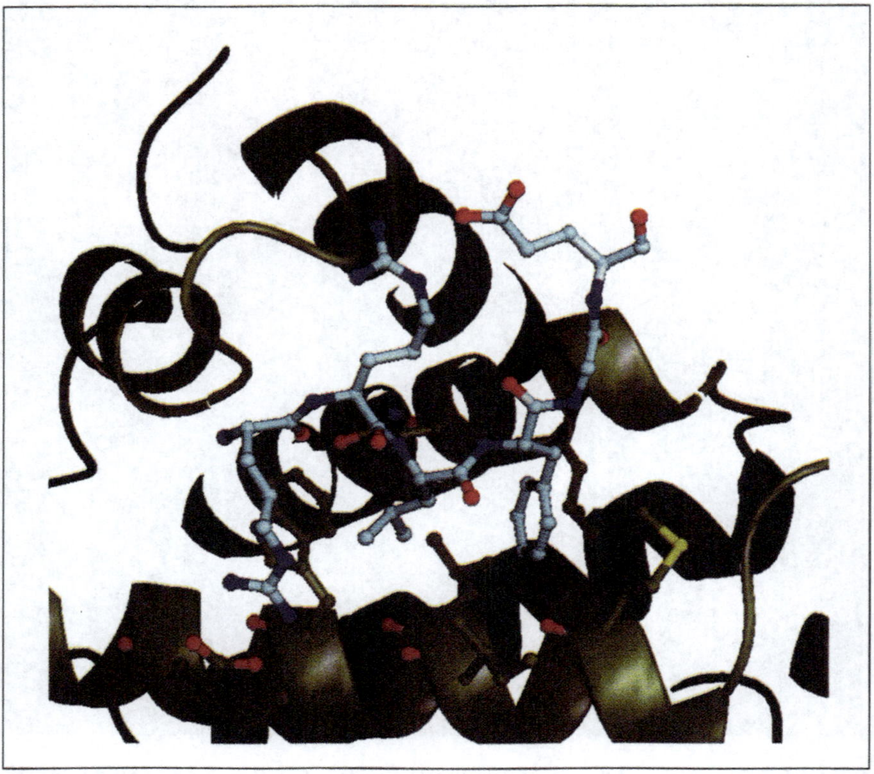

Figure C.4. The cyclin A substrate recruitment site as shown in Figure 6.4. The cyclin A3 structure in the vicinity of the substrate recruitment site is drawn in ribbon representation. The side-chains of selected residues that form the first helix of the N-terminal cyclin box fold are drawn in ball and stick reipresentation. A p107-derived recruitment peptide (sequence RRLFGE, backbone carbons in cyan) is bound into the hydrophobic ZRXL-binding pocket of cyclin A. This site has an important role in the binding of certain substrates and of Cip/Kip1 family CKIs to the CDK/cyclin complex and is conserved among A, B, D and E cyclins.

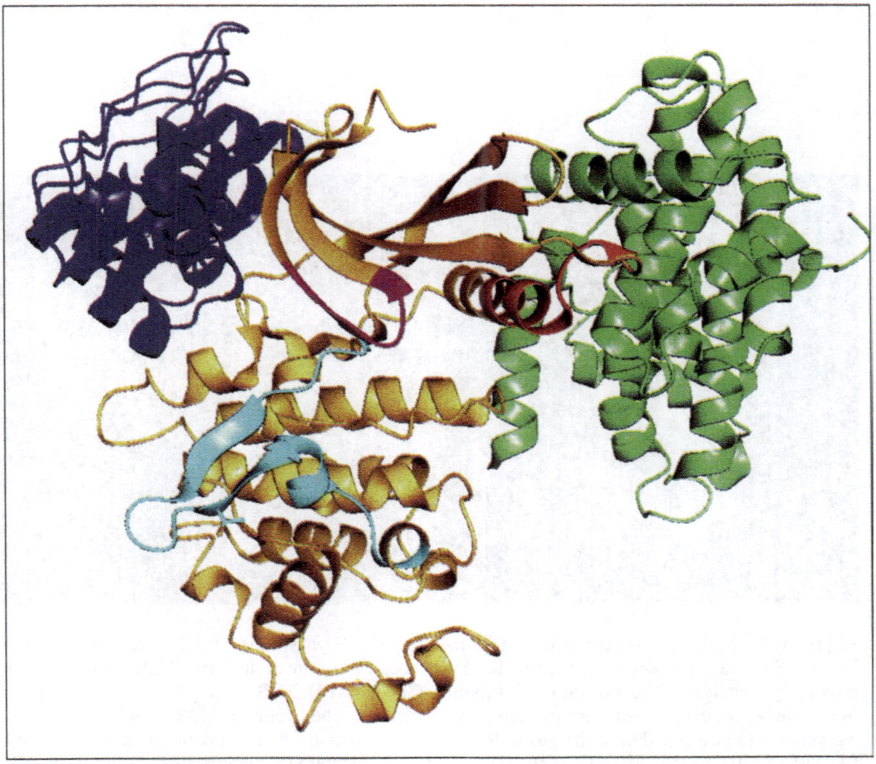

Figure C.5. Structure of CDK6 in complex with p18^{INK4c} and cyclin K as shown in Figure 6.5. The same view of the CDK6 structure as that selected for CDK2 in Figure 6.1 is shown. To aid comparison of the two CDK structures, the CDK subunit is again coloured gold, the cyclin sub-unit green and the CDK6 glycine loop (residues 20-25), activation loop (residues 163-189) and C-helix (residues 55-65) are drawn in pink, red and cyan respectively. The p18^{INK4c} structure is drawn in ribbon representation in purple. This view highlights the CDK6 activation loop structure and the displacement of the C helix from the active site cleft. The CDK6-cyclin K interface is much reduced in comparison with that observed in the CDK2/cyclin A complex (Fig. 6.1B). p18^{INK4c} interacts with the CDK6 N-terminal domain and does not contact cyclin K.

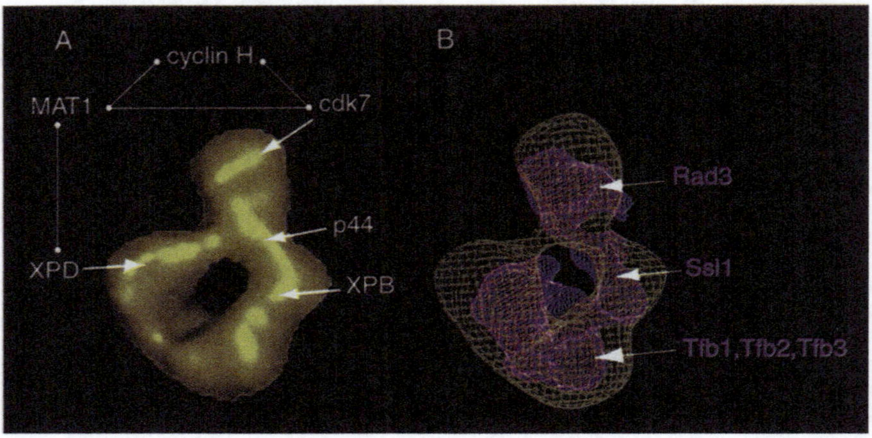

Figure C.6. Quaternary organizations of human TFIIH and yeast core TFIIH as shown in Figure 7.2. A) Surface representation of the 3D model reconstruction of human TFIIH from electron microscopy analysis. The positions of subunits CDK7, XPD XPB, and p44, as inferred from immunolabeling experiments, are indicated by the arrows. The cyclin H, MAT1, and p34 subunits are tentatively positioned according to well established pairwise subunit-subunit interaction studies. (Adapted from Chang et al, Cell 2000; 102:609-613.) B) Quaternary organizations of human TFIIH and yeast core TFIIH. Superimposition of the 2D crystal structure of yeast core TFIIH with human TFIIH (grey). Assignments of density regions to subunits in the yeast structure, either Rad3p, Ssl1p. or Tfb1p/Tfb2p/Tfb3p, are indicated. (Adapted from Chang et al, Cell 2000; 102:609-613.)

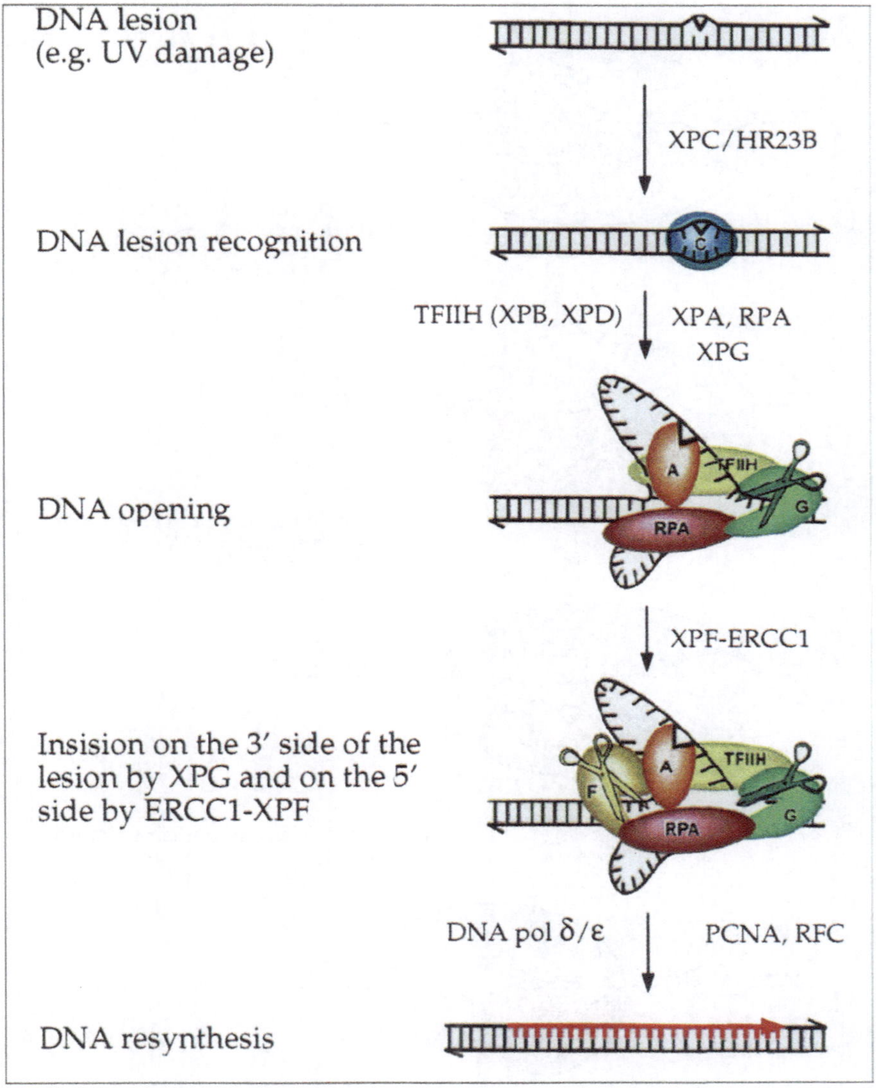

Figure C.7. Schematic representation of the nucleotide excision repair (NER) pathway as shown in figure 7.3. The involvement of the various NER proteins (XPC/HR23B, TFIIH, RPA, XPA, XPG, XPF-ERCC1, DNA polymerase, PCNA, RCF and DNA ligase 1) are shown at each step of the reaction (damage recognition, DNA opening, incision, excision, resysthesis and ligation), as described in the text.

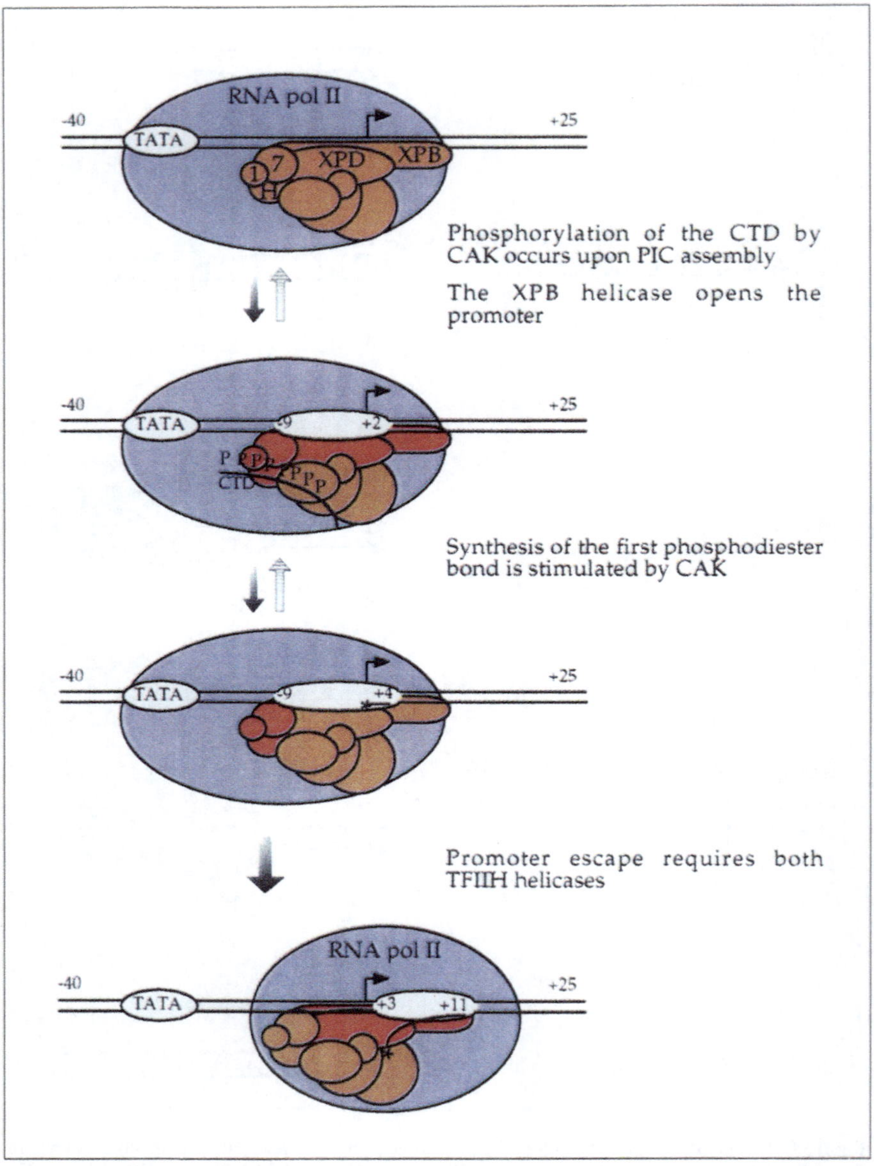

Figure C.8. Proposed role(s) of TFIIH subunits in transcription initiation as shown in Figure 7.4. The involvement of the CAK subcomplex or of the two helicases subunits of TFIIH into the different steps of the initiation of transcription are shown.

Chapter 1

CAK from Marine Invertebrates to Human

Randy Y.C. Poon

Introduction

The eukaryotic cell cycle is driven by an evolutionarily conserved engine consisting of a series of cyclin-dependent kinase (CDK)-cyclin modules. The orderly events of the cell cycle depend on the complex interplay of many factors, but the long-held idea that activation and inhibition of different cyclin-CDK complexes drive different phases of the cell cycle still remains a good approximation. The activities of CDKs are highly regulated both positively and negatively, and most of these regulators of the CDKs have been conserved down to molecular detail during evolution from yeast to human. However, it is interesting that CDK-activating kinase (CAK), the protein kinase that phosphorylates the conserved activating threonine residue in the CDKs, seem to have diverged considerably. The composition, functions, and regulation of CAK in higher eukaryotes are discussed in this review. CAK in yeast and other organisms are discussed elsewhere in this volume. Since there is no definite proof of the existence of other CAKs in metazoans at this moment, the complex containing cyclin H-CDK7-MAT1 will be presumed to be the CAK in metazoans here, although the evidence of the existence of other CAKs will also be discussed.

Regulation of the Cell Cycle by Cyclin-Dependent Kinases

Cyclins are defined as proteins that are related in sequence to the originally isolated mitotic A- and B-type cyclins. Cyclin-dependent kinases (CDKs) are defined as protein kinases whose kinase activities are dependent on binding to a cyclin subunit (see Table 1.1). All CDKs are related in sequence to the archetypal member CDC2 (with 40%-70% identity in protein sequence). In mammalian cells, cyclin B-CDC2 is the principal mitotic cyclin-CDK complex that regulates the G_2-M transition. Cyclin A can also associate with CDC2, and is synthesized and destroyed slightly earlier than cyclin B during G_2 phase. CDK2 associates with cyclin A and cyclin E (although it can also associate with cyclin D in fibroblasts), and the respective complexes are important regulators for S phase and G_1-S transition. CDK3 is very similar in sequence to CDK2, but its precise function is not known. CDK4 and CDK6 are partners of D-type cyclins that function in G_1 phase before cyclin E-CDK2.

It has become apparent that several cyclins and CDKs have non-cell cycle regulatory functions. For instance, cyclin D was also found to be able to bind and activate the estrogen receptor. CDK5, first identified as a partner of cyclin D in fibroblasts, is activated in postmitotic neurons by binding to a protein called p35, which shares no sequence homology to cyclin. CDK7, which is a component of CDK-activating kinase (CAK), is also a component of the basal transcription factor TFIIH and can phosphorylate the carboxy-terminal repeat domain (CTD) of the large subunit of RNA polymerase II (see later). Other cyclins and CDKs (cyclin

The CDK-Activating Kinase (CAK), edited by Philipp Kaldis.

Table 1.1. The human CDK family

CDK	T-loop sequence	Cyclin partner	Function
CDC2	154-GIPIR--------VYTHEVVTLW-168	A, B	G_2-M
CDK2	153-GVPVR------TYTHEVVTLW-167	A, D, E	S, G_1-S
CDK3	153-GVPLR-------TYTHEVVTLW-167	E?	G_1-S?
CDK4	166-SYQM--------ALTPVVVTLW-179	D	G_1
CDK5	152-GIPVR--------CYSAEVVTLW-166	D, p35, p39	neuronal
CDK6	171-SFQM---------ALTSVVVTLW-184	D	G_1
CDK7	163-GSPNR-------AYTHQVVTRW-177	H	CAK,transcription
CDK8	181-NSPLKP---LADLDPVVVTFW-198	C	transcription
CDK9	175-SLAKNSQPNRYTNRVVTLW-193	T	transcription

The CDK family, their cyclin partners and the known functions are indicated. Note that the sequences of the regulatory subunit of CDK5 in neuronal cells, p35 and p39, do not correspond to cyclin. An alignment of the protein sequences around the activating residue at the T-loop region is shown. The threonine residues that are phosphorylated by CAK in CDC2, CDK2, CDK4 and CDK6 are underlined. The corresponding serine/threonine residues in other CDKs, as well as other residues that are phosphorylated by other protein kinases are also highlighted.

C-CDK8, cyclin T-CDK9, and cyclin K) also associate with RNA polymerase II and can phosphorylate CTD.

Regulation of CDKs in Higher Eukaryotic Cells

The activity of CDK is, by definition, dependent on binding to a cyclin subunit. Cyclin expression is controlled both at the transcriptional and post-translational levels. The post-translational regulation of cyclins occurs mainly through ubiquitin-mediated degradation, carried out by mechanisms involving the anaphase-promoting complex (APC) (or the cyclosome) and the SCF (SKP1-Cullin-F box protein) complexes. Phosphorylation of a threonine residue (Thr161 in CDC2) on the T-loop activates CDKs, but phosphorylation of a threonine and a tyrosine residue (Thr14 and Tyr15 in CDC2) close to the catalytic core inhibits the activity of CDKs. The Thr14 and Tyr15 residues of CDC2 can be phosphorylated by the protein kinases MYT1 and WEE1 respectively, and dephosphorylated by members of the CDC25 phosphatase family (A, B and C). CDKs can also be inhibited by two families of CDK inhibitors: the p16 family ($p16^{INK4A}$, $p15^{INK4B}$, $p18^{INK4C}$, and $p19^{INK4D}$), which is specific for cyclin D-CDK4/6 complexes, and the p21 family ($p21^{CIP1/WAF1}$, $p27^{KIP1}$, $p57^{KIP2}$), which can inhibit a broader spectrum of CDKs. Finally, formation of some active cyclin-CDK complexes may involve assembly factors and chaperones like MAT1 and CDC37.

The Role of T-loop Phosphorylation in CDK Activation

Phosphorylation of a specific threonine residue in the activation loop of CDKs (Thr161 in CDC2, and Thr160 in CDK2) is essential for maximal CDK activation. Bacterially expressed cyclin A can bind to CDK2 and result in a $4x10^5$-fold activation of the kinase, but phosphorylation of Thr160 by CAK activates cyclin A-CDK2 a further 80-fold (1). Mutation of the Thr161/Thr160 in CDC2/CDK2 to a nonphosphorylatable alanine abolished the maximal activation of the CDK.[1,2] These biochemical analyses are in good agreement with the crystal structures of cyclin A-CDK2 complexes. Monomeric CDK2 has an overall bi-lobed structure typical of serine/threonine protein kinases, but there are several deviations that explain why the activation of CDK2 requires cyclin binding and Thr160 phosphorylation. In comparison to the structures of other active protein kinases, the ATP in the catalytic core of monomeric CDK2 is not aligned correctly for phospho-transfer, and the T-loop containing the Thr160 residue blocks the access of potential substrates to the catalytic site.[3] Upon binding to cyclin A, the T-loop is largely moved away from blocking the catalytic site.[4] Finally, upon phosphorylation of Thr160 by CAK, the T-loop moves a further ~7 Å away and the cyclin A-CDK2 contact is increased.[5]

The change in cyclin-CDK interaction after Thr161/Thr160 phosphorylation may explain the observation that while the formation of some cyclin-CDK pairs (cyclin A-CDK2, cyclin B-CDC2, cyclin D-CDK4, cyclin E-CDK2) is not affected by CAK phosphorylation, the formation of cyclin A-CDC2 is stimulated by CAK phosphorylation.[6-8] It is interesting that substituting a serine residue for Thr160 in CDK2 does not affect activation by cyclin A and CAK (in fact, Ser160 is more efficiently phosphorylated by CAK than Thr160).[9] However, dephosphorylation of Ser160 by HeLa cell extract or purified PP2Cb is slower than with Thr160.

Monomeric CDKs are poor substrates for CAK, but binding of the cyclin subunit stimulates the phosphorylation by CAK.[10-13] In the crystal structure of cyclin A-CDK2, this can be seen as making the T-loop more accessible for CAK phosphorylation in comparison to monomeric CDK2.[5] Interestingly, a single mutation in cyclin D1 (residue threonine 156 mutated to an alanine or a glutamic acid) allows cyclin D1 to form a complex with CDK4, but CDK4 cannot be phosphorylated by CAK.[14] The requirement of cyclin binding for CAK to phosphorylate the CDK may explain the cell cycle variation of Thr161/Thr160 phosphorylation.[15] This could also be due to the activity of the phosphatase that dephosphorylates the Thr161/Thr160. The CDK-interacting dual-specificity phosphatase KAP can dephosphorylate CDK2 Thr160.[16] KAP only dephosphorylates Thr160 in monomeric CDK2 after the destruction of the cyclin A partner. Type 2C protein phosphatases (PP2Cs) have also been shown to dephosphorylate yeast Cdc28p and human CDK2 in HeLa cell extracts.[17] It was shown that the majority of CDK2 phosphatase activity in cell-free extracts is Mg^{2+}-dependent and not inhibited by the KAP inhibitor vanadate. The crystal structure of KAP in association with Thr160-phosphorylated CDK2 may explain these data (David Barford, personal communication). It was found that KAP is readily inactivated by oxidation because of the formation of a disulphite bond between the catalytic site Cys residue and a neighbouring Cys residue. One possibility is that KAP present in cell-free extracts is easily inactivated by oxidation.

Activation of CDKs by Cyclin H-CDK7-MAT1 Complexes

CAK was first defined in *Xenopus* egg extracts biochemically as an enzyme that can phosphorylate and activate cyclin-CDC2.[18] The catalytic subunit of CAK was identified in starfish oocytes and *Xenopus* egg extracts as a protein kinase previously characterized as MO15.[10,12,19] These early purifications and characterizations of CAK have already suggested that like the CDKs, activation of MO15 may require phosphorylation and association with a companion subunit. The regulatory subunit of CAK was later identified as a cyclin (cyclin H) by protein purification and yeast two-hybrid screening.[11,20] Given that MO15 is a cyclin H-dependent kinase, it was later renamed CDK7. It was a striking revelation that a cyclin-CDK pair is involved in the activation of other cyclin-CDK complexes. In contrast to other cyclin-CDKs, cyclin H-CDK7 binds to an additional subunit called MAT1 (ménage à trois).[21-24] MAT1 contains a highly conserved C3HC4 zinc-binding domain, or RING finger, at the N-terminal region. However, binding of MAT1 to cyclin H-CDK7 involves the C-terminus of MAT1 and does not require the RING finger domain.[23,25]

CAK was initially characterized as a CDC2/CDK2-activating kinase, but the phosphorylation of other CDKs by CAK has also been investigated. Upon binding to D-type cyclins, CDK4 and CDK6 can be phosphorylated at the threonine residues on the T loop (Thr172 and Thr177 respectively) by CAK.[8,13,26,27] Cyclin D can also form a complex with CDK2, but the cyclin D-CDK2 complexes cannot be phosphorylated by CAK and are not active.[28] The precise function of CDK3 is not known, but it can also be phosphorylated by CAK in vitro.[29] Although CDK5 contains a serine residue (Ser159) at the equivalent position of Thr161 in CDC2, CDK5 is not phosphorylated or activated by CAK.[30,31] However, it was demonstrated that CDK5 could be phosphorylated and activated by an activity in PC12 cell lysates, and mutation of Ser159 to alanine abolished this activation.[31] This suggests that CDK5 can be activated by a protein kinase distinct from CAK. In this connection, it was shown that casein kinase I can phosphorylate and activate CDK5 in vitro.[31] The protein sequences around Thr161 or equivalent residues in the CDKs are shown in Table 1.1.

Regulation of the Enzymatic Activity of CAK

Like other CDKs, the activity of CDK7 itself is also regulated by multiple mechanisms. MAT1 activates cyclin H-CDK7 by acting both as an assembly factor, as well as substituting for the activating phosphorylation in CDK7.[24] MAT1 is not the only example of the involvement of assembly factors or chaperones in the formation of cyclin-CDK complexes. Mammalian CDC37 and the heat shock protein HSP90 were found to associate with CDK4, and to a lesser extent CDK6 in vitro and in vivo.[32-34] In yeast, the G_1 arrest in a *CDC37* temperature sensitive mutant strain is accompanied by a decrease in the complex formation between Cln2p and Cdc28p,[35] indicating that Cdc37p may have a role in regulating cyclin-CDK assembly. However, definite proof that CDC37 is an assembly factor for cyclin D-CDK4 is lacking. Unlike the case of cyclin H-CDK7-MAT1, no cyclin D is present in the CDK4-CDC37-HSP90 complex. CDK4-CDC37-HSP90 association is likely to involve the newly synthesized monomeric CDK4 and may assist the proper folding of the CDK4, since inhibition of HSP90 function with geldanamycin reduces the level of CDK4 and decreases the stability of newly synthesized CDK4.[34] The p21$^{CIP1/WAF1}$ class of CDK inhibitors have also been reported to have a role in promoting the formation of cyclin D1-CDK4 complexes.[36,37]

What is interesting about MAT1 is that it promotes the assembly of cyclin H-CDK7, as well as bypassing the requirement for phosphorylation in the activation of CAK.[24] Two residues on the activating loop region of CDK7 (Ser164 and Thr170 in human CDK7) are phosphorylated in the cell. Phosphorylation of Ser164 (a CDC2 consensus phosphorylation site) is dispensable for cyclin H-CDK7 activity, but phosphorylation of Thr170 (the equivalent site to Thr161 in CDC2) is required for cyclin H-CDK7 activity.[29,38] Although CDK7 is itself a CDK, the Ser164 and Thr170 residues are not autophosphorylated in CAK. This suggests the presence of another CAK-activating kinase (CAKAK). Intriguingly, Thr170 in cyclin H-CDK7 can be phosphorylated by targets of CAK, such as cyclin A-CDK2 and cyclin B-CDC2, in an autocatalytic loop.[24]

Phosphorylation of the activating loop appears to stimulate CDK7 binding to cyclin H (unlike the majority of other cyclin-CDK complexes).[24,39] Both Ser164 and Thr170 can be phosphorylated by cyclin A-CDK2 and cyclin B-CDC2 in the absence of cyclin H and MAT1. Phosphorylation of Ser164 is sufficient for CDK7 to bind to cyclin H with low affinity, but high affinity binding requires Thr170 phosphorylation.[39] Furthermore, phosphorylation of both Ser164 and Thr170 together confers significant CAK activity on CDK7 even in the absence of cyclin H and MAT1.[24,39] Indeed, active CAK containing heterodimeric cyclin H-CDK7 without MAT1 can be found in *Xenopus* oocytes.[39] Other studies reveal a more complex relationship between Ser164 phosphorylation and CDK7 activity.[40] Phosphorylation of Ser164 is increased during mitosis in yeast, and this phosphorylation inhibits the TFIIH-associated kinase and transcription activities (see below). Taken together, there appears to be alternative pathways to activate CAK, and there is a complex interplay between the two phosphorylation sites in CDK7, MAT1 binding, and cyclin H binding on CAK activation. Finally, MAT1 is also implicated in substrate recognition for CDK7 (see below).

Regulation of CAK in the Cell

Given that phosphorylation of the T-loop threonine is absolutely required for the maximal activation of CDKs, it is conceivable that variation of CAK activity may contribute to the timing of CDK activation. Therefore it was somewhat surprising to find that both the level and activity of CAK remain constant in the cell cycle. The mRNA of CDK7 remains relatively constant through the cell cycle,[41] and CDK7 is a stable protein with a half-life > 16 hr.[42] The levels of CDK7 and CAK activity are constant during early embryonic cell cycle[38,42] and somatic cell cycle.[27,38,43,44] The association of the three components of CAK is also invariant throughout the cell cycle.[44] In cells that have exited the cell cycle in states like quiescence or senescence in fibroblasts, the level of CAK has been reported to be either similar[27,45] or reduced.[38,46,47]

Why the requirement of phosphorylation of Thr161 has evolved at all when the CAK activity is not regulated? One possibility is that although the level and activity of CAK are constant during the cell cycle, the accessibility for the CDKs may be regulated. CDK7 is localized to the nucleus.[21,43-45] The KRKR nuclear localization sequence located at the C-terminal end of CDK7 is required for nuclear targeting.[29,43] Interestingly, nuclear targeting of CDK7 is necessary to produce active CAK, because CDK7 with a mutated nuclear localization sequence is unable to generate CAK as long as the nuclear envelope is not broken in *Xenopus* oocytes.[43] Further investigations show that cyclin H, CAK7, MAT1, as well as other components of the TFIIH are localized to the subnuclear structures known as coiled bodies (or Cajal bodies).[48] The nuclear localization of CAK is consistent with the localization of its substrates CDK2, CDK4, and CDK6. Indeed, cyclin E-CDK2 (but not cyclin A) is also found to be localized to the coiled bodies.[49] CDK2 is not enriched in the coiled bodies compartment unless it is complexed with cyclin E, suggesting that cyclin E may recruit CDK2 to the coiled bodies to be phosphorylated by CAK. However, the major cyclin-CDK complex for G_2-M, cyclin B-CDC2, is localized to the cytoplasm.[50] How CDC2 can be phosphorylated by CAK during the G_2 phase represents one of the problems in considering cyclin H-CDK7-MAT1 as the bona fide CAK in mammalian cells (see below).

Apart from the possibility of regulation by subcellular compartmentation, phosphorylation of CDKs by CAK can also be regulated by the CDK inhibitors. It has been shown that for a variety of conditions that lead to cell cycle arrest, the $p21^{CIP1/WAF1}$ class of CDK inhibitors are induced and can block the phosphorylation of CDKs by CAK. In macrophages, cAMP blocks the mitogenic effects of colony-stimulating factor-1 by induction of $p27^{KIP1}$.[8] $p27^{KIP1}$ inhibits the kinase activity of cyclin D-CDK4 either directly, or by blocking the phosphorylation of CDK4 by CAK without directly affecting the activity of CAK. Similarly, it was demonstrated that $p21^{CIP1/WAF1}$ or $p27^{KIP1}$ induced by interferon,[51] prostaglandin A2,[52] contact inhibition,[53] or replicative senescence[54] inhibits the phosphorylation of CDKs by CAK. Phosphorylation of CDKs by CAK is blocked by both the $p21^{CIP1/WAF1}$ class and the $p16^{INK1A}$ class of CDK inhibitors.[13,26] The N-terminal minimal CDK binding domain of $p27^{KIP1}$ (amino acids 28-81) is sufficient to inhibit the phosphorylation of cyclin E-CDK2 by CAK.[55] The blockage of CAK phosphorylation by $p27^{KIP1}$ is not that surprising in view of the relatively large structural disruption of the N-terminal lobe of CDK2 caused by $p27^{KIP1}$.[56] Similarly, large conformational changes of the catalytic cleft and the cyclin-binding site are induced in CDK6 after binding to the $p16^{INK4A}$ family of inhibitors.[57,58]

Other Substrates of Mammalian CAK

Substrates other than CDKs can also be phosphorylated by mammalian CAK. These substrates mainly composed of proteins of the transcription machinery and transcription factors. The cyclin H-CDK7-MAT1 complex is a component of the basal RNA polymerase II transcription factor TFIIH. Cyclin H-CDK7 can phosphorylate the carboxy-terminal repeat domain (CTD) of the large subunit of RNA polymerase II, which is important for transcriptional initiation and nucleotide-excision repair.[21,59-63] Indeed, transcription of some promoters in vitro requires the activity of CDK7.[40,64] Association of MAT1 or TFIIH with cyclin H-CDK7 switched its substrate preference to favor CTD over CDK2.[65,66]

Apart from cyclin H-CDK7, other cyclin-CDK complexes are also found to associate with RNA polymerase II and phosphorylate the CTD. Cyclin C represents a case where functional complementation does not always reveal a protein's true character. Despite the fact that cyclin C was originally cloned by complementation of the yeast G_1 cyclins *CLN1-3*, there is no indication that cyclin C is involved in G_1 regulation. Instead, cyclin C-CDK8 associates with the large subunit of RNA polymerase II and can phosphorylate CTD.[67,68] Another cyclin C-related cyclin called cyclin T interacts directly with the human immunodeficiency virus type 1 Tat protein.[69] Cyclin T binds to CDK9 and has CTD kinase and CAK activities.[70,71] Interaction of Tat with cyclin T-CDK9 enhances the binding of Tat and the positive transcription elongation factor (P-TEFb) to the viral TAR RNA stem-loop structure, which is important to promote transcription elongation and overcome transcriptional pausing induced by the TAR element. Autophosphorylation of cyclin T-CDK9 complexes is required for high-affinity binding of Tat to TAR RNA.[72] Tat also binds to cyclin H-CDK7 and stimulates the phosphorylation of CTD.[71,73] But there are different suggestions as to whether the cyclin H-CDK7 association with TFIIH is important[74] or dispensable[75] for Tat function. Another cyclin, cyclin K, also binds to RNA polymerase II and associates with both CTD kinase and CAK activities.[76] It has been suggested that although several cyclin-CDK complexes can phosphorylate CTD, they may phosphorylate different residues within CTD.[77]

Other substrates that have been shown to be phosphorylated by CAK include TFIIE and TFIIF,[65,66] retinoic acid receptor RARα,[78] estrogen receptor ERα,[79] octamer binding transcription factors,[80] and the tumor suppressor p53.[81,82] Phosphorylation of these transcription factors by CAK tends to modulate their activities. A single CAK phosphorylation site has been reported in RARα and ERα, but residues at both the N-terminal[81] and C-terminal regions[82] of p53 have been reported to be phosphorylated by CAK. The CAK phosphorylation sites in different substrates identified so far show little conservation in the primary sequences, suggesting that CAK has a very loose primary structure requirement.

Evidence of the Existence of Other CAKs in Higher Eukaryotes

Although cyclin H-CDK7-MAT1 accounts for the majority of CAK activity in mammalian cells and *Xenopus* eggs,[83] there are data that pose a serious challenge to the physiological relevance of this complex in higher eukaryotic cells.

(a) The active CDK7-containing complex is in the nucleus, whereas one of its presumed substrate cyclin B-CDC2 is in the cytoplasm. However, evidence for the constant shuttling of cyclin B1 between the nucleus and the cytoplasm may obviate this problem.[84,85]

(b) The cyclin H-CDK7-MAT1 complex is a component of TFIIH and can phosphorylate CTD (see above). Both free and TFIIH-associated cyclin H-CDK7-MAT1 are present in the cell. The free complex has higher specificity towards CDKs and the TFIIH-complex has higher specificity towards CTD.[65,66,86] This apparently dichotomous function of cyclin

H-CDK7 in CAK and TFIIH is problematic because the budding yeast version of cyclin H-CDK7, Ccl1p-Kin28p, has CTD kinase activity but not CAK activity in vitro.[87,88]

(c) Some cyclins and CDKs, like cyclin T-CDK9 and cyclin K, also contain CAK activity in vitro.

(d) CAK identified from yeast does not resemble the mammalian cyclin H-CDK7-MAT1 complex.[89-91] The *S. cerevisiae* CAK is a single subunit 43 kDa protein kinase (called Cak1p or Civ1p) that shares limited sequence similarity to the CDK family, and is fully active as a monomer. *CAK1* is an essential gene, and is responsible for the phosphorylation of several CDKs (including Cdc28p and Kin28p, but not Pho85p and Srb10p) but not the CTD in yeast.[92]

Since Ccl1p-Kin28p and cyclin H-CDK7 appear to have different substrate specificity in vitro, and both Cak1p and cyclin H-CDK7 seem to account for the majority of CAK activity in the respective organism, we are left with the fascinating possibility that CAK and TFIIH CTD kinase are distinct enzymes in yeast, whereas cyclin H-CDK7 performs both functions in higher eukaryotes. This idea is supported by experiments done in *Xenopus* egg extracts, in which immunodepletion of CDK7 suppressed CAK activity and inhibited entry into M phase.[83] CAK activity was restored by adding back CDK7 and its associated subunits, demonstrating that a CDK7 complex is necessary and sufficient for activation of cyclin-CDK in *Xenopus* eggs. Stronger support comes from the demonstration that CDK7 is essential for the activation of cyclin A/B-CDC2 complexes and for mitosis in *Drosophila*.[93] However, activation of CDK2 does not appear to be compromised in *Drosophila* with mutations in *CDK7*. Whether an additional Cak1p-like CAK is present in metazoans is still a contentious issue. It does not help to clear up the issue when a CAK unlike either CDK7 or Cak1p has been identified in *Arabidopsis thaliana*.[94] Moreover, both a cyclin H-CDK7-like pair (Mcs2-Mcs6) and an unrelated kinase Csk1 have been identified to have CAK activity in *S. pombe*.[95] Interestingly, Csk1 can also function as a CAKAK for Mcs2-Mcs6 in vivo.[96] In *Xenopus* egg extracts and mammalian cell extracts, CAK activity was seen as a single peak over several purification steps.[10,11] But in subsequent characterizations of CAK activities in HeLa cells using CDK-affinity chromatography, a second CAK activity was detected that runs on gel filtration at 30-40 kDa.[97] This second CAK activity cross-reacts with antibodies against the yeast Cak1p but not CDK7, and resembles Cak1p rather than CDK7 in terms of substrate specificity and sensitivity to an ATP analog. This protein that cross-reacts with the antibodies against yeast Cak1p is apparently down-regulated in TGFβ-treated HepG2 cells.[98]

Conclusion

The cyclin H-CDK7-MAT1 complex is the best-characterized candidate for CAK in higher eukaryotic cells, which is entirely unlike the CAK in budding yeast. CAK phosphorylates the threonine residue on the T-loop in CDC2, CDK2, CDK3, CDK4, and CDK6 in mammalian cells. Unlike typical CDKs, activation of CDK7 involves complex interplay between phosphorylation of two residues on the T-loop and association with the assembly factor MAT1. The activity of CAK is not regulated in the mammalian cell cycle, but phosphorylation of CDKs by CAK may be regulated by subcellular localization, cyclins, and the CDK inhibitors. Cyclin H-CDK7-MAT1 is also a component of the basal RNA polymerase II transcription factor TFIIH, and can phosphorylate the CTD of the large subunit of RNA polymerase II and several transcription factors. Whether cyclin H-CDK7-MAT1 is the bona fide CAK in metazoans is still a contentious issue.

Acknowledgments

I apologize for those whose work that could not be cited due to space constraints. Many thanks are due to members of my laboratory for comments. Works in my laboratory are supported by the Biotechnology Research Institute and the Research Grants Council.

References

1. Connell-Crowley L, Solomon MJ, Wei N, Harper JW. Phosphorylation independent activation of human cyclin-dependent kinase 2 by cyclin A in vitro. Mol Biol Cell 1993; 4(1):79-92.
2. Solomon MJ, Lee T, Kirschner MW. Role of phosphorylation in $p34^{cdc2}$ activation: Identification of an activating kinase. Mol Biol Cell 1992; 3(1):13-27.
3. De Bondt HL, Rosenblatt J, Jancarik J, Jones HD et al. Crystal structure of human CDK2: Implications for the regulation of cyclin-dependent kinases by phosphorylation and cyclin binding. Nature 1993; 363:595-602.
4. Jeffrey P, Russo AA, Polyak K, Gibbs E et al. Mechanism of CDK activation revealed by the structure of a cyclin A-CDK2 complex. Nature 1995; 376:313-320.
5. Russo AA, Jeffrey PD, Pavletich NP. Structural basis of cyclin-dependent kinase activation by phosphorylation. Nature Structural Biol. 1996; 3:696-700.
6. Desai D, Wessling HC, Fisher RP, Morgan DO. Effects of phosphorylation by CAK on cyclin binding by CDC2 and CDK2. Mol Cell Biol 1995; 15(1):345-350.
7. Brown NR, Noble ME, Lawrie AM, Morris MC, Tunnah P et al. Effects of phosphorylation of threonine 160 on cyclin-dependent kinase 2 structure and activity. J Biol Chem 1999; 274(13):8746-8756.
8. Kato JY, Matsuoka M, Strom DK, Sherr CJ. Regulation of cyclin D-dependent kinase 4 (cdk4) by cdk4-activating kinase. Mol Cell Biol 1994; 14(4):2713-2721.
9. Kaldis P, Cheng A, Solomon MJ. The effects of changing the site of activating phosphorylation in cdk2 from threonine to serine. J Biol Chem 2000; 275(42):32578-32584.
10. Solomon MJ, Harper JW, Shuttleworth J. CAK, the $p34^{cdc2}$ activating kinase, contains a protein identical or closely related to $p40^{MO15}$. EMBO J 1993; 12(8):3133-3142.
11. Fisher RP, Morgan DO. A novel cyclin associates with MO15/CDK7 to form the CDK-activating kinase. Cell 1994; 78(4):713-724.
12. Fesquet D, Labbe JC, Derancourt J, Capony JP, Galas S et al. The MO15 gene encodes the catalytic subunit of a protein kinase that activates cdc2 and other cyclin-dependent kinases (CDKs) through phosphorylation of Thr161 and its homologues. EMBO J 1993; 12(8):3111-3121.
13. Kaldis P, Russo AA, Chou HS, Pavletich NP, Solomon MJ. Human and yeast cdk-activating kinases (CAKs) display distinct substrate specificities. Mol Biol Cell 1998; 9(9):2545-2560.
14. Diehl JA, Sherr CJ. A dominant-negative cyclin D1 mutant prevents nuclear import of cyclin-dependent kinase 4 (CDK4) and its phosphorylation by CDK-activating kinase. Mol Cell Biol 1997; 17(12):7362-7374.
15. Krek W, Nigg EA. Differential phosphorylation of vertebrate $p34^{cdc2}$ kinase at the G1/S and G2/M transitions of the cell cycle: Identification of major phosphorylation sites. EMBO J 1991; 10(2):305-316.
16. Poon RYC, Hunter T. Dephosphorylation of Cdk2 Thr160 by the cyclin-dependent kinase-interacting phosphatase KAP in the absence of cyclin. Science 1995; 270(5233):90-93.
17. Cheng A, Ross KE, Kaldis P, Solomon MJ. Dephosphorylation of cyclin-dependent kinases by type 2C protein phosphatases. Genes Dev 1999; 13(22):2946-2957.
18. Solomon MJ. The function(s) of CAK, the $p34^{cdc2}$-activating kinase. Trends Biochem Sci 1994; 19(11):496-500.
19. Poon RYC, Yamashita K, Adamczewski JP, Hunt T, Shuttleworth J. The cdc2-related protein $p40^{MO15}$ is the catalytic subunit of a protein kinase that can activate $p33^{cdk2}$ and $p34^{cdc2}$. EMBO J 1993; 12(8):3123-3132.
20. Makela TP, Tassan JP, Nigg EA, Frutiger S, Hughes GJ et al. A cyclin associated with the CDK-activating kinase MO15. Nature 1994; 371(6494):254-257.
21. Tassan JP, Jaquenoud M, Fry AM, Frutiger S, Hughes GJ et al. In vitro assembly of a functional human CDK7-cyclin H complex requires MAT1, a novel 36 kDa RING finger protein. EMBO J 1995; 14(22):5608-5617.
22. Yee A, Nichols MA, Wu L, Hall FL, Kobayashi R et al. Molecular cloning of CDK7-associated human MAT1, a cyclin-dependent kinase-activating kinase (CAK) assembly factor. Cancer Res 1995; 55(24):6058-6062.

23. Devault A, Martinez AM, Fesquet D, Labbe JC, Morin N et al. MAT1 ('menage a trois') a new RING finger protein subunit stabilizing cyclin H-cdk7 complexes in starfish and Xenopus CAK. EMBO J 1995; 14(20):5027-5036.
24. Fisher RP, Jin P, Chamberlin HM, Morgan DO. Alternative mechanisms of CAK assembly require an assembly factor or an activating kinase. Cell 1995; 83(1):47-57.
25. Busso D, Keriel A, Sandrock B, Poterszman A, Gileadi O et al. Distinct regions of MAT1 regulate cdk7 kinase and TFIIH transcription activities. J Biol Chem 2000; 275(30):22815-22823.
26. Aprelikova O, Xiong Y, Liu ET. Both p16 and p21 families of cyclin-dependent kinase (CDK) inhibitors block the phosphorylation of cyclin-dependent kinases by the CDK- activating kinase. J Biol Chem 1995; 270(31):18195-18197.
27. Matsuoka M, Kato JY, Fisher RP, Morgan DO, Sherr CJ. Activation of cyclin-dependent kinase 4 (cdk4) by mouse MO15-associated kinase. Mol Cell Biol 1994; 14(11):7265-7275.
28. Higashi H, Suzuki-Takahashi I, Saitoh S, Segawa K, Taya Y et al. Cyclin-dependent kinase-2 (Cdk2) forms an inactive complex with cyclin D1 since Cdk2 associated with cyclin D1 is not phosphorylated by Cdk7- cyclin-H. Eur J Biochem 1996; 237(2):460-467.
29. Labbe JC, Martinez AM, Fesquet D, Capony JP, Darbon JM et al. $p40^{MO15}$ associates with a p36 subunit and requires both nuclear translocation and Thr176 phosphorylation to generate cdk-activating kinase activity in Xenopus oocytes. EMBO J 1994; 13(21):5155-5164.
30. Poon RYC, Lew J, Hunter T. Identification of functional domains in the neuronal Cdk5 activator protein. J Biol Chem 1997; 272(9):5703-5708.
31. Sharma P, Sharma M, Amin ND, Albers RW, Pant HC. Regulation of cyclin-dependent kinase 5 catalytic activity by phosphorylation. Proc Natl Acad Sci USA 1999; 96(20):11156-11160.
32. Dai K, Kobayashi R, Beach D. Physical interaction of mammalian CDC37 with CDK4. J Biol Chem 1996; 271:22030-22034.
33. Lamphere L, Fiore F, Xu X, Brizuela L, Keezer S, Sardet C et al. Interaction between Cdc37 and Cdk4 in human cells. Oncogene 1997; 14(16):1999-2004.
34. Stepanova L, Leng X, Parker SB, Harper JW. Mammalian $p50^{Cdc37}$ is a protein kinase-targeting subunit of Hsp90 that binds and stabilizes Cdk4. Genes Dev 1996; 10:1491-1502.
35. Gerber MR, Farrell A, Deshaies RJ, Herskowitz I, Morgan DO. Cdc37 is required for association of the protein kinase Cdc28 with G1 and mitotic cyclins. Proc Natl Acad Sci USA 1995; 92(10):4651-4655.
36. Cheng M, Olivier P, Diehl JA, Fero M, Roussel MF, Roberts JM et al. The $p21^{(Cip1)}$ and $p27^{(Kip1)}$ CDK 'inhibitors' are essential activators of cyclin D-dependent kinases in murine fibroblasts. EMBO J 1999; 18(6):1571-1583.
37. LaBaer J, Garrett MD, Stevenson LF, Slingerland JM, Sandhu C, Chou HS et al. New functional activities for the p21 family of CDK inhibitors. Genes Dev 1997; 11(7):847-862.
38. Poon RYC, Yamashita K, Howell M, Ershler MA, Belyavsky A et al. Cell cycle regulation of the $p34^{cdc2}/p33^{cdk2}$-activating kinase $p40^{MO15}$. J Cell Sci 1994; 107(10):2789-2799.
39. Martinez AM, Afshar M, Martin F, Cavadore JC, Labbe JC, Doree M. Dual phosphorylation of the T-loop in cdk7: its role in controlling cyclin H binding and CAK activity. EMBO J 1997; 16(2):343-354.
40. Akoulitchev S, Reinberg D. The molecular mechanism of mitotic inhibition of TFIIH is mediated by phosphorylation of CDK7. Genes Dev 1998; 12(22):3541-3550.
41. Levedakou EN, He M, Baptist EW, Craven RJ, Cance WG, Welcsh PL et al. Two novel human serine/threonine kinases with homologies to the cell cycle regulating Xenopus MO15, and NIMA kinases: Cloning and characterization of their expression pattern. Oncogene 1994; 9(7):1977-1988.
42. Brown AJ, Jones T, Shuttleworth J. Expression and activity of $p40^{MO15}$, the catalytic subunit of cdk-activating kinase, during *Xenopus* oogenesis and embryogenesis. Mol Biol Cell 1994; 5(8):921-932.
43. Darbon JM, Devault A, Taviaux S, Fesquet D, Martinez AM, Galas S et al. Cloning, expression and subcellular localization of the human homolog of $p40^{MO15}$ catalytic subunit of cdk-activating kinase. Oncogene 1994; 9(11):3127-3138.

44. Tassan JP, Schultz SJ, Bartek J, Nigg EA. Cell cycle analysis of the activity, subcellular localization, and subunit composition of human CAK (CDK-activating kinase). J Cell Biol 1994; 127(2):467-478.
45. Bartkova J, Zemanova M, Bartek J. Expression of CDK7/CAK in normal and tumor cells of diverse histogenesis, cell-cycle position and differentiation. Int J Cancer 1996; 66(6):732-737.
46. Hamada K, Takuwa N, Zhou W, Kumada M, Takuwa Y. Protein kinase C inhibits the CAK-CDK2 cyclin-dependent kinase cascade and G1/S cell cycle progression in human diploid fibroblasts. Biochim Biophys Acta 1996; 1310(1):149-156.
47. Wong H, Riabowol K. Differential CDK-inhibitor gene expression in aging human diploid fibroblasts. Exp Gerontol 1996; 31(1-2):311-325.
48. Jordan P, Cunha C, Carmo-Fonseca M. The cdk7-cyclin H-MAT1 complex associated with TFIIH is localized in coiled bodies. Mol Biol Cell 1997; 8(7):1207-1217.
49. Liu J, Hebert MD, Ye Y, Templeton DJ, Kung H, Matera AG. Cell cycle-dependent localization of the CDK2-cyclin E complex in Cajal (coiled) bodies. J Cell Sci 2000; 113(9):1543-1552.
50. Pines J, Hunter T. Human cyclins A and B1 are differentially located in the cell and undergo cell cycle-dependent nuclear transport. J Cell Biol 1991; 115(1):1-17.
51. Mandal M, Bandyopadhyay D, Goepfert TM, Kumar R. Interferon-induces expression of cyclin-dependent kinase-inhibitors $p21^{WAF1}$ and $p27^{Kip1}$ that prevent activation of cyclin-dependent kinase by CDK-activating kinase (CAK). Oncogene 1998; 16(2):217-225.
52. Hitomi M, Shu J, Agarwal M, Agarwal A, Stacey DW. $p21^{Waf1}$ inhibits the activity of cyclin dependent kinase 2 by preventing its activating phosphorylation. Oncogene 1998; 17(8):959-969.
53. Kato A, Takahashi H, Takahashi Y, Matsushime H. Inactivation of the cyclin D-dependent kinase in the rat fibroblast cell line, 3Y1, induced by contact inhibition. J Biol Chem 1997; 272(12):8065-8070.
54. Morisaki H, Ando A, Nagata Y, Pereira-Smith O, Smith JR et al. Complex mechanisms underlying impaired activation of Cdk4 and Cdk2 in replicative senescence: Roles of p16, p21, and cyclin D1. Exp Cell Res 1999; 253(2):503-510.
55. Rank KB, Evans DB, Sharma SK. The N-terminal domains of cyclin-dependent kinase inhibitory proteins block the phosphorylation of cdk2/Cyclin E by the CDK-activating kinase. Biochem Biophys Res Commun 2000; 271(2):469-473.
56. Russo AA, Jeffrey PD, Patten AK, Massagué J, Pavletich NP. Crystal structure of the $p27^{Kip1}$ cyclin-dependent-kinase inhibitor bound to the cyclin A-Cdk2 complex. Nature 1996; 382:325-331.
57. Brotherton DH, Dhanaraj V, Wick S, Brizuela L, Domaille PJ et al. Crystal structure of the complex of the cyclin D-dependent kinase Cdk6 bound to the cell-cycle inhibitor $p19^{INK4d}$. Nature 1998; 395(6699):244-250.
58. Russo AA, Tong L, Lee JO, Jeffrey PD, Pavletich NP. Structural basis for inhibition of the cyclin-dependent kinase Cdk6 by the tumour suppressor $p16^{INK4a}$. Nature 1998; 395(6699):237-243.
59. Feaver WJ, Svejstrup JQ, Henry NL, Kornberg RD. Relationship of CDK-activating kinase and RNA polymerase II CTD kinase TFIIH/TFIIK. Cell 1994; 79(6):1103-1109.
60. Serizawa H, Makela TP, Conaway JW, Conaway RC, Weinberg RA et al. Association of Cdk-activating kinase subunits with transcription factor TFIIH. Nature 1995; 374(6519):280-282.
61. Roy R, Adamczewski JP, Seroz T, Vermeulen W, Tassan JP, Schaeffer L et al. The MO15 cell cycle kinase is associated with the TFIIH transcription-DNA repair factor. Cell 1994; 79(6):1093-1101.
62. Shiekhattar R, Mermelstein F, Fisher RP, Drapkin R, Dynlacht B et al. Cdk-activating kinase complex is a component of human transcription factor TFIIH. Nature 1995; 374(6519):283-287.
63. Adamczewski JP, Rossignol M, Tassan JP, Nigg EA, Moncollin V et al. MAT1, cdk7 and cyclin H form a kinase complex which is UV light- sensitive upon association with TFIIH. EMBO J 1996; 15(8):1877-1884.

64. Mäkelä TP, Parvin JD, Kim J, Huber LJ, Sharp PA et al. A kinase-deficient transcription factor TFIIH is functional in basal and activated transcription. Proc Natl Acad Sci USA 1995; 92(11):5174-5178.
65. Yankulov KY, Bentley DL. Regulation of CDK7 substrate specificity by MAT1 and TFIIH. EMBO J 1997; 16(7):1638-1646.
66. Rossignol M, Kolb-Cheynel I, Egly JM. Substrate specificity of the cdk-activating kinase (CAK) is altered upon association with TFIIH. EMBO J 1997; 16(7):1628-1637.
67. Leclerc V, Tassan JP, O'Farrell PH, Nigg EA, Leopold P. Drosophila Cdk8, a kinase partner of cyclin C that interacts with the large subunit of RNA polymerase II. Mol Biol Cell 1996; 7(4):505-513.
68. Rickert P, Seghezzi W, Shanahan F, Cho H, Lees E. Cyclin C/CDK8 is a novel CTD kinase associated with RNA polymerase II. Oncogene 1996; 12:2631-2640.
69. Wei P, Garber ME, Fang SM, Fischer WH, Jones KA. A novel CDK9-associated C-type cyclin interacts directly with HIV-1 Tat and mediates its high-affinity, loop-specific binding to TAR RNA. Cell 1998; 92(4):451-462.
70. Yang X, Gold MO, Tang DN, Lewis DE, Aguilar-Cordova E, Rice AP et al. TAK, an HIV Tat-associated kinase, is a member of the cyclin-dependent family of protein kinases and is induced by activation of peripheral blood lymphocytes and differentiation of promonocytic cell lines. Proc Nat Acad Sci USA 1997; 94(23):12331-12336.
71. Nekhai S, Shukla RR, Kumar A. A human primary T-lymphocyte-derived human immunodeficiency virus type 1 Tat-associated kinase phosphorylates the C-terminal domain of RNA polymerase II and induces CAK activity. J Virol 1997; 71(10):7436-7441.
72. Garber ME, Mayall TP, Suess EM, Meisenhelder J, Thompson NE et al. CDK9 autophosphorylation regulates high-affinity binding of the human immunodeficiency virus type 1 tat-P-TEFb complex to TAR RNA. Mol Cell Biol 2000; 20(18):6958-6969.
73. Cujec TP, Okamoto H, Fujinaga K, Meyer J, Chamberlin H, Morgan DO, et al. The HIV transactivator TAT binds to the CDK-activating kinase and activates the phosphorylation of the carboxy-terminal domain of RNA polymerase II. Genes Dev 1997; 11(20):2645-2657.
74. Okamoto H, Cujec TP, Peterlin BM, Okamoto T. HIV-1 replication is inhibited by a pseudo-substrate peptide that blocks Tat transactivation. Virology 2000; 270(2):337-344.
75. Chen D, Zhou Q. Tat activates human immunodeficiency virus type 1 transcriptional elongation independent of TFIIH kinase. Mol Cell Biol 1999; 19(4):2863-2871.
76. Edwards MC, Wong C, Elledge SJ. Human cyclin K, a novel RNA polymerase II-associated cyclin possessing both carboxy-terminal domain kinase and Cdk-activating kinase activity. Mol Cell Biol 1998; 18(7):4291-4300.
77. Rickert P, Corden JL, Lees E. Cyclin C/CDK8 and cyclin H/CDK7/p36 are biochemically distinct CTD kinases. Oncogene 1999; 18(4):1093-1102.
78. Rochette-Egly C, Adam S, Rossignol M, Egly JM, Chambon P. Stimulation of RAR alpha activation function AF-1 through binding to the general transcription factor TFIIH and phosphorylation by CDK7. Cell 1997; 90(1):97-107.
79. Chen D, Riedl T, Washbrook E, Pace PE, Coombes RC et al. Activation of estrogen receptor alpha by S118 phosphorylation involves a ligand-dependent interaction with TFIIH and participation of CDK7. Mol Cell 2000; 6(1):127-137.
80. Inamoto S, Segil N, Pan ZQ, Kimura M, Roeder RG. The cyclin-dependent kinase-activating kinase (CAK) assembly factor, MAT1, targets and enhances CAK activity on the POU domains of octamer transcription factors. J Biol Chem 1997; 272(47):29852-29858.
81. Ko LJ, Shieh SY, Chen X, Jayaraman L, Tamai K et al. p53 is phosphorylated by CDK7-cyclin H in a $p36^{MAT1}$-dependent manner. Mol Cell Biol 1997; 17(12):7220-7229.
82. Lu H, Fisher RP, Bailey P, Levine AJ. The CDK7-cycH-p36 complex of transcription factor IIH phosphorylates p53, enhancing its sequence-specific DNA binding activity in vitro. Mol Cell Biol 1997; 17(10):5923-5934.
83. Fesquet D, Morin N, Doree M, Devault A. Is Cdk7/cyclin H/MAT1 the genuine cdk activating kinase in cycling Xenopus egg extracts? Oncogene 1997; 15(11):1303-1307.
84. Hagting A, Karlsson C, Clute P, Jackman M, Pines J. MPF localization is controlled by nuclear export. EMBO J 1998; 17(14):4127-4138.

85. Toyoshima F, Moriguchi T, Wada A, Fukuda M, Nishida E. Nuclear export of cyclin B1 and its possible role in the DNA damage- induced G2 checkpoint. EMBO J 1998; 17(10):2728-2735.
86. Drapkin R, Le Roy G, Cho H, Akoulitchev S, Reinberg D. Human cyclin-dependent kinase-activating kinase exists in three distinct complexes. Proc Natl Acad Sci USA 1996; 93(13):6488-6493.
87. Cismowski MJ, Laff GM, Solomon MJ, Reed SI. KIN28 encodes a C-terminal domain kinase that controls mRNA transcription in Saccharomyces cerevisiae but lacks cyclin-dependent kinase-activating kinase (CAK) activity. Mol Cell Biol 1995; 15(6):2983-2992.
88. Valay JG, Simon M, Dubois MF, Bensaude O, Facca C, Faye G. The KIN28 gene is required both for RNA polymerase II mediated transcription and phosphorylation of the Rpb1p CTD. J Mol Biol 1995; 249(3):535-544.
89. Espinoza FH, Farrell A, Erdjument-Bromage H, Tempst P, Morgan DO. A cyclin-dependent kinase-activating kinase (CAK) in budding yeast unrelated to vertebrate CAK. Science 1996; 273(5282):1714-1717.
90. Kaldis P, Sutton A, Solomon MJ. The Cdk-activating kinase (CAK) from budding yeast. Cell 1996; 86(4):553-564.
91. Thuret JY, Valay JG, Faye G, Mann C. Civ1 (CAK in vivo), a novel Cdk-activating kinase. Cell 1996; 86(4):565-576.
92. Espinoza FH, Farrell A, Nourse JL, Chamberlin HM, Gileadi O, Morgan DO. Cak1 is required for Kin28 phosphorylation and activation in vivo [published erratum appears in Mol Cell Biol 2000 Mar; 20(5):1898]. Mol Cell Biol 1998; 18(11):6365-6373.
93. Larochelle S, Pandur J, Fisher RP, Salz HK, Suter B. Cdk7 is essential for mitosis and for in vivo Cdk-activating kinase activity. Genes Dev 1998; 12(3):370-381.
94. Umeda M, Bhalerao RP, Schell J, Uchimiya H, Koncz C. A distinct cyclin-dependent kinase-activating kinase of Arabidopsis thaliana. Proc Natl Acad Sci USA 1998; 95(9):5021-5026.
95. Lee KM, Saiz JE, Barton WA, Fisher RP. Cdc2 activation in fission yeast depends on Mcs6 and Csk1, two partially redundant Cdk-activating kinases (CAKs). Curr Biol 1999; 9(8):441-444.
96. Hermand D, Pihlak A, Westerling T, Damagnez V, Vandenhaute J, Cottarel G, et al. Fission yeast Csk1 is a CAK-activating kinase (CAKAK). EMBO J 1998; 17(24):7230-7238.
97. Kaldis P, Solomon MJ. Analysis of CAK activities from human cells. Eur J Biochem 2000; 267(13):4213-4221.
98. Nagahara H, Ezhevsky SA, Vocero-Akbani AM, Kaldis P, Solomon MJ et al. Transforming growth factor beta targeted inactivation of cyclin E:cyclin-dependent kinase 2 (Cdk2) complexes by inhibition of Cdk2 activating kinase activity. Proc Natl Acad Sci USA 1999; 96(26):14961-14966.

CHAPTER 2

Activating Phosphorylation of Cyclin-Dependent Kinases in Budding Yeast

Philipp Kaldis, Vasiliki Tsakraklides, Karen E. Ross, Edward Winter and Aiyang Cheng

Introduction

The eukaryotic cell division cycle is controlled via the sequential activation and inactivation of cyclin-dependent protein kinases (CDKs). In the budding yeast *Saccharomyces cerevisiae*, Cdc28p is the only CDK involved in regulating the cell cycle, while in higher eukaryotes, multiple CDKs (CDC2 [CDK1], CDK2, CDK4, and CDK6) control cell cycle progression. CDK substrates are assumed to be key players in most major cell cycle events. To ensure the proper timing and coordination of cell cycle events, intracellular and extracellular signals modulate CDK activities through a variety of mechanisms, including association with regulatory subunits (cyclins, inhibitors, and assembly factors), subcellular localization, transcriptional regulation, selective proteolysis, and reversible protein phosphorylation.[1-7]

Binding of cyclins to CDKs is required for kinase activity. Cyclins also likely contribute to the substrate specificity[8,9] and to the subcellular localization of CDKs.[10-14] Furthermore, cyclin binding leads to multiple phosphorylations on the CDK in *Xenopus* egg extracts.[15,16] Cyclins are synthesized, and degraded periodically during the cell cycle.[1,4,17,18] The synthesis of cyclins is mostly controlled at the level of transcription[19] and they are degraded by ubiquitin-mediated proteolysis. Cyclins are recognized by the ubiquitin-protein ligases APC (anaphase-promoting complex) or SCF (Skp1p-Cdc53p/cullin-F-box protein), polyubiquitinated, and destroyed by the 26S proteasome.[20,21] In the budding yeast, nine cyclins form complexes with Cdc28p: three G1 cyclins (Cln1p to Cln3p) and six B-type cyclins (Clb1p to Clb6p). In the G1 phase, Cln1-3p regulate cell cycle events such as "Start" (equivalent of restriction point in mammalian cells),[22,23] the repression of pheromone-induced transcription, and meiosis in diploid cells.[24,25] B-type cyclins are expressed in three successive waves from Start to mitosis and regulate most events during the cell cycle (reviewed in 26).

In budding yeast, there are two Cdc28p inhibitors (CKIs): Far1p and Sic1p. Far1p, which is required for pheromone-induced cell cycle arrest,[27] inhibits the kinase activities of Cln1p-Cdc28p, Cln2p-Cdc28p, and Cln3p-Cdc28p,[22,28,29] but it is incapable of reducing Clb5p or Clb2p associated Cdc28p activities.[29] In contrast to Far1p, Sic1p is a specific inhibitor of B-type cyclin-Cdc28p complexes.[30,31] Sic1p has two functions: in G1, it prevents premature S-phase entry before bud initiation and spindle pole body (SPD) duplication.[31] At the end of

The CDK-Activating Kinase (CAK), edited by Philipp Kaldis.

mitosis, it promotes mitotic exit by inhibiting Clb-Cdc28p activity.[32,33] It should be noted that there is no apparent sequence homology between yeast CKIs and their mammalian counterparts. Mammalian CKIs are classified in two groups: the CIP/KIP family consisting of $p21^{CIP1/WAF1}$,[34-36] $p27^{KIP1}$,[37-39] and $p57^{KIP2}$[40,41] which inhibit all CDKs; and the INK4 family including $p15^{INK4b}$,[42] $p16^{INK4a}$,[43] $p18^{INK4c}$[44] and $p19^{INK4d}$[45,46] that are specific CDK4/CDK6 inhibitors.

In addition to binding of protein subunits, CDKs are regulated by protein phosphorylation (for review see 7). There are two inhibitory and one activating phosphorylation sites. The inhibitory phosphorylations (which occur on Thr-14 and Tyr-15 in human CDC2) are carried out by the WEE1-like protein kinases (Swe1p in budding yeast), and removed by the CDC25 phosphatase family (Mih1p in budding yeast) (for reviews see 3, 7). In higher eukaryotes, inhibitory phosphorylation regulates timing of entry into M phase.[47] In *Saccharomyces cerevisiae*, the inhibitory phosphorylation of Cdc28p on Tyr-19 by Swe1p does not play a significant role during a normal cell cycle or after DNA damage[48,49] but is required for a morphologic checkpoint that monitors the coordination of the budding and nuclear division cycles.[50] Activating phosphorylation occurs within the so-called T-loop[7,51] on a conserved threonine residue corresponding to Thr-169 in Cdc28p and Thr-161 in human CDC2. The inhibitory phosphorylations are dominant over the activating phosphorylation. Full activation of CDKs, which is necessary for normal cell cycle progression, requires binding of a cyclin, removal of inhibitory phosphorylations, and the presence of an activating phosphorylation. The latter is required for CDK activities in vitro and in vivo[16,52-57] and is catalyzed by the CDK-activating kinase (CAK).[58] In higher eukaryotes, CAK was identified as the CDK7/cyclin H/MAT1 complex (see Chapter 1). In this review, we will focus on the regulation of Cdc28p and other yeast CDKs by Cak1p, the budding yeast *Saccharomyces cerevisiae* CAK and by the protein phosphatase 2C (PP2C).

Historical View of Cak1p Identification

After CAK activity was detected in cell extracts, it was purified from *Xenopus* and starfish egg extracts and later from human and mouse cell extracts (see Chapter 1). However, determination of the physiological role of CAK is difficult in mammalian cells due to technical limitations. Therefore, it was of great interest to identify CAK in a genetically tractable system like the budding yeast. Three groups, including our own, succeeded in isolating the protein responsible for CAK in budding yeast by biochemical methods. An assumption that we and others made in these studies, was that the CAK activity in budding yeast extracts would be able to activate recombinant human CDK2/cyclin complexes. It is worth noting that none of the groups identifying Cak1p used Cdc28p as substrate since it is difficult to express, it tends to aggregate, and displays erratic low activity. Human CDK2 was the preferred assay substrate in these studies since it can be easily expressed and purified. Furthermore, CDK2 has been shown to complement Cdc28p mutations[59] indicating that it can be a Cak1p substrate in vivo. In the CAK activity assays, inactive CDK2/cyclin complexes were incubated with yeast cell extracts, thereby CDK2 could be activated and its ability to phosphorylate histone H1 was assayed. Otherwise the assay was similar to one described earlier for studying CAK from frog and human extracts.[56]

In mammals, CDK7 was isolated as an enzyme with CAK activity (see Chapter 1) and was later shown to be a subunit of TFIIH (see Chapter 7). The closest homolog to CDK7 in the budding yeast is Kin28p. Kin28p was originally cloned as a Cdc28p-like gene[60] that has a cyclin-like regulatory subunit, Ccl1p.[61] The only function known for Kin28p was that it affected the transcription of some genes.[62,63] The correlation to CDK7 became almost perfect when Kin28p was shown to be a subunit of yeast TFIIH.[64] Surprisingly, Kin28p does not display any CAK activity nor does it affect the phosphorylation state of Cdc28p.[62,64] Because of the discrepancy between CDK7 and Kin28p it was worthwhile to purify the protein responsible for CAK activity in budding yeast.

Purification of the CDK2 activating activity led to the identification of a 43 kDa protein which was named Cak1p[52,54] or Civ1p.[57] The successful purification of Cak1p depended on several critical factors including the activity assay and the ability to prepare large amounts of biochemically active yeast lysates. The source of budding yeast was another major concern in these studies. One group grew large amounts of yeast in a fermenter and lysed them by mechanical disruption ["bead beating"].[52] Our group used freeze dried yeast (Sigma #YCS-2) because these lysates displayed robust CAK activity. In addition, we lysed the yeast by grinding them under liquid nitrogen which allows the preparation of large amounts of lysate.[54] In the end, both groups purified the same protein confirming the suitability of both approaches.[52,54]

Thuret et al[57] used a different biochemical approach involving the isolation of Cdc28p from yeast extracts. They realized that the addition of cyclin A readily activated Cdc28p and hypothesized that CAK was part of the purified Cdc28p complex. After gel filtration of Cdc28p complexes, they found a CAK activity peak at 45 kDa and identified it as Cak1p (also called Civ1p) using a candidate-gene approach. They also found Cak1p in Cdc28p immunoprecipitates. Nevertheless, it is curious that such a Cak1p/Cdc28p complex of 80 kDa or more has never been observed after gel filtration[52,54,57] and Cdc28p does not interact with Cak1p in a two-hybrid screen.[65]

Four different genetic screens yielded the *CAK1* gene, which was called at the time *CAK1*,[66] *PDL3*,[67] *MCA28*[63,68] or *MEB1*[69] [see below]. The first screen involved the protein phosphatase Sit4p that is required for progression from G1 to S phase.[70,71] In the strain background used in these studies, absence of Sit4p causes a slow growth phenotype. *CAK1* (formerly known as *PDL3*) was identified as being synthetically lethal with *sit4* mutations.[67] The relation between *CAK1* and *SIT4* is unclear and might be indirect. The second screen used a yeast haploid-lethal mutant collection that was examined microscopically to identify mutants arrested with a morphology similar to that caused by mutation in *CDC4*, *CDC34*, or *CDC53*, genes that are subunits of the SCF complex.[69] One mutant was found to be identical to *CAK1*. Again the relation between *CAK1* and *CDC4, 34, 53* is not obvious, except that all these genes regulate the activity of Cdc28p through degradation of cyclins and inhibitors. In the third screen, *CAK1* was isolated as a dosage suppressor of the sporulation defect of a *smk1-2* conditional strain.[66] *SMK1* is a MAP kinase involved in spore wall morphogenesis during meiosis in yeast. The functions of *CAK1* in meiosis and its activity in the Smk1 pathway are discussed in further detail below, see "Cak1p and its role in meiotic development." In the fourth screen, Cak1p (formerly named *MCA28*) was identified as a dosage-dependent suppressor of ***kin28-ts*** mutations.[63,68] Kin28p is a subunit of transcription factor IIH where it phosphorylates the C-terminal domain (CTD) of the large subunit of RNA polymerase II[63,64] [for more details on Kin28p see section on Substrates and Biochemistry of Cak1p].

Interestingly, none of the genetic screens revealed the function of *CAK1* in Cdc28p activation. The reason for this might be that *CAK1* is involved in all phases of the cell cycle and therefore genetically interacts with many gene products leading to a complex mutant phenotype from which it is difficult to infer Cak1's function. A reverse approach like biochemical purification which followed function to sequence proved most successful in identifying Cak1p and its roles in the yeast cell cycle.

Although CDK7 homologs have been identified in many species from fission yeast to humans, only a few potential Cak1p-like enzymes have been identified. Csk1 of *S. pombe*, Cak1At of *Arabidopsis thaliana*, and Cak1p of *Candida albicans* have each been shown to rescue a *cak1* mutant in *S. cerevisiae*.[72-74] Cak1At is closer in primary amino acid sequence to CDK7 whereas the Csk1 sequence resembles that of Cak1p. Studies in fission yeast indicate that Csk1 indeed acts as a monomer and can phosphorylate the CAK Mcs6 in vivo and cdc2 in vitro suggesting that is has CAKAK (CAK activating kinase) as well as CAK activity.[73,75] Nevertheless, in vivo Csk1 only activates Mcs6 but not cdc2 [for details see Chapter 3].[76] Whereas CDK7 possesses both CAK and CTD kinase activities, neither Csk1 nor Cak1At, have been found to phosphorylate the CTD.[73,74] We have indications that Cak1At, Csk1, and *C. albicans* Cak1p expressed and purified in the insect cell system are functional as a monomer and prefer to phosphorylate monomeric CDKs (authors' unpublished observations). Cyclin subunits have not been described for any of these enzymes. Thus these proteins appear to follow the pattern of budding yeast Cak1p rather than the heterotrimeric CDK7 metazoan model.

Substrates and Biochemistry of Cak1p

Cak1p belongs to the family of serine/threonine protein kinases and is a distant member of the CDK superfamily.[52,54,57] Cak1p activates CDKs in vitro by phosphorylation of the conserved CDK domain known as the T-loop [also called activation segment].[77] Cak1p activates the yeast cell cycle CDK Cdc28p[52,54,57] as well as its functional homologs *Xenopus* Cdc2,[78] human CDK3 (P. K. unpublished data) and human CDK2.[52,54,57] Cak1p also phosphorylates CDK6 in its monomeric as well as cyclin D1-, D2-, or D3- bound forms.[77] The metazoan CAK CDK7/cyclin H phosphorylates CDC2, CDK2, CDK3, CDK4, and the CDK6/cyclin D3 complex but not CDK6/cyclin D1/2 complexes,[77] a distinction in substrate utilization that opens the possibility for the existence of CDK activating kinases tailored for specific CDK. Cak1p does not phosphorylate all CDKs. In the same activity assays that determined its activity on CDK6 for example, no kinase activity was observed with CDK4.[77] The K_M(CDK2) of Cak1p increases by a factor of 4 when the activating T-loop threonine is mutated to a serine [T160S].[79] Such an effect was not observed for CDK7, once again underlining the difference in the biochemistry of the two CAKs.[79] It is interesting to note that phosphorylation of CDK2 by CAK does not serve to stabilize ATP in the ATPase reaction but increases the catalytic efficiency by 100,000-fold and the turnover rate by 1,000-fold.[80]

An important finding of in vitro activity assays was that Cak1p demonstrates a substrate preference for monomeric CDK rather than the cyclin-bound form.[77] This is in contrast to the mode of action of CDK7, which preferentially phosphorylates CDK/cyclin complexes.[77,81] The functional implications of this selectivity on the Cdc28p activation pathway will be discussed later in this review. It is important to note the mechanistic significance of the diverged substrate specificity between Cak1p and CDK7. Both CAKs can phosphorylate CDK2 yet binding of cyclin oppositely influences their activity, suggesting that each might recognize CDK2 in a slightly different conformation (given that cyclin binding induces conformational changes in CDK structure).

In vivo, Cak1p's primary substrate is the major cell cycle CDK Cdc28p. In fact, phosphorylation of Cdc28p at Thr169 (the CDK T-loop site of activating phosphorylation) is the only essential function of Cak1p for normal budding yeast growth.[52,54,57] A strain that carries a mutant version of Cdc28-43244p mimicking constitutive phosphorylation (T169E) and harboring several additional mutations in the same gene, bypasses the requirement for activating phosphorylation and is viable independently of Cak1p.[82] Furthermore, this mutant Cdc28-43244p is capable of supporting cell cycle progression, showing that regulated reversible phosphorylation at position Thr-169 is not essential under the conditions tested. Although viable, the strain has growth defects in the absence of Cak1p perhaps because of failure to phosphorylate other, as yet unknown, nonessential substrates of Cak1p.[82] Conspicuous in its absence from a list of Cak1p substrates is the C-terminal domain (CTD) of RNA polymerase II's large subunit. CDK7 is a subunit of the transcription factor TFIIH complex and phosphorylates the CTD.[64,83-86] Phosphorylation of the CTD by TFIIH is thought to mediate the transition from transcription initiation to elongation. Cak1p was neither found in the TFIIH complex nor does it exhibit CTD kinase activity.[52,54,57]

Such activity however is found in CDK7's closest *S. cerevisiae* relative, the cyclin-dependent kinase Kin28p[62] which is part of the yeast TFIIH complex.[64] Cak1p may yet maintain a connection to transcription as it was found to phosphorylate Kin28p at the conserved threonine (Thr-162) and stimulate its CTD kinase activity in vitro. Kin28p remains unphosphorylated when Cak1p is inactivated in vivo indicating that it is a physiological Cak1p substrate.[87,88] This phosphorylation event, however, is not essential as a T162A mutant of Kin28p is viable.[88] In the absence of Thr-162 phosphorylation, Kin28p can be activated instead through binding to the assembly factor Tbf3p/Rig2p.[89,90]

Biochemical Characterization of Cak1p

Cak1p is unconventional in that its primary amino acid sequence deviates in subdomains shared by the protein kinase superfamily. For example, budding yeast Cak1p entirely lacks a canonical "glycine loop" (GxGxxG) in the nucleotide-binding fold. Typically, this motif is instrumental in anchoring the nucleotide substrate and its mutation in PKA leads to reduced catalytic activity.[91] Cak1p however maintains reasonable affinity for ATP [K_M(ATP) = 4.8 μM, ATPase rate = 0.13 min^{-1}, compared to 17 μM and 0.66 min^{-1} for wild-type PKA].[92] It can also bind GTP (though with a K_M over 200-fold higher than for ATP) as well as ADP and AMPPNP.[92] Introduction of a glycine motif by mutagenesis did not significantly affect the enzyme's catalytic rate suggesting that other subdomains in Cak1p may be atypical.[92] Cak1p has substitutions in highly conserved kinase core residues, and although it contains the "invariant lysine" essential for activity in most protein kinases, this residue is dispensable for Cak1p activity.[92] In other kinases, the invariant lysine is involved in nucleotide alignment and seems to function in catalysis as its mutation has little effect on nucleotide binding but completely inhibits catalysis.[93] Mutations of the Cak1p Lys-31 surprisingly reduced ATP binding but left catalytic activity largely unaffected. These mutants were able to complement a *cak1* deletion in vivo suggesting that this residue is not required for activity.[66,67,69,92] Additionally, Cak1p is insensitive to 5'-fluorosulfonylbenzoyladenosine [FSBA][77,92] which through covalent modification of the invariant lysine leads to loss of activity in nearly all kinases, including CDK7.[56] Thus a combination of kinetic and mutagenic analysis demonstrate that Cak1p's unusual ATP binding pocket is efficiently functional. It remains to be resolved what structural adaptations provide the alternative route by which the molecule compensates for the lack of canonical protein kinase features. In an effort to further understand Cak1p biochemistry, Enke et al[94] examined the parameters of CDK2 phosphorylation by Cak1p. It was concluded that catalysis is the rate limiting step in CDK2 phosphorylation which likely proceeds through a rapid-equilibrium, random pathway (where nucleotide and protein substrate binding are independent in the formation of the ternary complex). Despite sequence similarity to the CDK family, Cak1p is not cyclin dependent. Purified Cak1p is active as a monomer in vitro and is not found associated with a cyclin or any other protein in vivo as it elutes in a single peak corresponding to its molecular weight when yeast extracts are subjected to gel filtration.[52,54,57,95] Furthermore, CDKs bound to mammalian CDK inhibitors (CKI; $p21^{CIP1}$, $p27^{KIP1}$, $p57^{KIP2}$, $p16^{INK4a}$, and $p18^{INK4c}$) were phosphorylated by Cak1p but not by human CDK7.[77] Crystallographic evidence[96] suggest that CKIs inhibit CDKs by binding to the substrate, by inducing conformational changes, and potentially sterically hindering enzyme-substrate interaction. Cak1p's insensitivity to CKIs suggests that Cak1p approaches substrates in a manner that is structurally distinct from that of CDK7 or that it recognizes CDKs in a different conformation. However, we should keep in mind that also CAKs might induce a conformational change when they bind to their CDK substrate.

Regulation and Localization of Cak1p

Regulation of CDKs plays an important role in their function. Since the binding of cyclins and the activating phosphorylation are general requirements for most of the CDKs including CDK7-type CAK, it was thought that budding yeast Cak1p might also be regulated by a similar mechanism. Surprisingly, Cak1p is active without a cyclin partner and does not require post-translational modification for its full kinase activity.[54] Furthermore, Cak1p in crude yeast extract elutes as a monomer from a gel filtration column. Mutation of potential phosphorylation sites in Cak1p has no effect on its activity in vivo or in vitro.[95] In addition, isoelectric focusing of Cak1p reveals no modified species, suggesting that the majority of Cak1p is not phosphorylated.[95] The activity and protein level of Cak1p remain constant during vegetative cell growth.[52,67] In contrast, *CAK1* mRNA[66] and protein levels[95] fluctuate dramatically during meiosis (discussed in section on Cak1p and its Roles in Meiotic Development). Although Cak1p protein levels and activity are constant, we cannot exclude that Cdc28p can only be activated at a specific time. In this model, both Cak1p and Cdc28p are present at all times but Cdc28p can only be phosphorylated by Cak1p at a specific time. That way Cak1p would regulate Cdc28p directly. Nevertheless, we have no indication that such a model is valid since Cdc28p is fully phosphorylated at all times [see section on The Activation Pathway of Cdc28p].[97] We cannot exclude though that only a small percentage of Cdc28p is phosphorylated at specific times or that this modification is only short lived. In both cases, it would be virtually impossible to detect such effects.

Whereas CDK7 is localized to the nucleus, budding yeast Cak1p is found in both the cytoplasm and the nucleus by immunofluorescence staining and is mostly found in the cytoplasmic fraction by subcellular fractionation,[95] which correlates with the localization of its major substrate, Cdc28p.[98] Furthermore, the protein levels and localization of Cak1p are constant throughout the cell cycle,[52,67] which is also consistent with the invariable phosphorylation state of Cdc28p.[97] It is unclear whether the activating phosphorylation of Cdc28p plays any regulatory role during the cell cycle in yeast.

Cak1p and its Roles in Meiotic Development

Sporulation is the program in which diploid (MATa/MATα) yeast cells give rise to haploid spores in response to nutrient limitation (reviewed in 99). Following induction, a single round of DNA replication is followed by an elongated prophase when synapsis and genetic recombination take place. Once recombination has been completed, two rounds of chromosome segregation occur without an intervening S-phase. Homologs separate during the unique MI division, while sister chromatids separate during the mitosis-like MII division. Spore wall assembly follows the meiotic divisions. Cak1p is required for multiple processes during sporulation. Some Cak1p requirements during sporulation appear to be Cdc28p-dependent while others appear to be Cdc28p-independent. In order to review the roles of Cak1p in meiotic development it is first necessary to consider Cdc28p's role in sporulation and how the regulation of Cdc28p may differ between mitosis and meiosis.

The role of Cdc28p in meiosis was first addressed by Shuster and Byers who reported that a *cdc28-4* diploid strain shifted to its restrictive temperature following meiotic induction completed chromosomal DNA replication and spindle pole body duplication and arrested prior to the nuclear divisions in pachytene [the last stage of MI prophase before cells become committed to chromosome segregation].[100] *CDC28* is required for both DNA synthesis and spindle pole body duplication during mitosis. In seeming contradiction to the *CDC28* temperature shift studies, premeiotic DNA replication does require the *CLB5* and *CLB6* B-type cyclins .[101, 102] Taken together, these results suggest that there is a low threshold of Cdc28p catalytic activity required to complete premeiotic DNA synthesis which is provided by the residual activity of the *cdc28-4* allele. Another possibility that cannot be ruled out is that there is an as yet to be identified CDK, that complexes with Clb5p and Clb6p and promotes DNA replication in meiosis. Premeiotic and mitotic DNA replication also differ in how CDK inhibitors are regulated. In mitosis, S-phase requires targeting of the Sic1p CDK inhibitor for destruction by Cdc28p/Cln1p and Cdc28p/Cln2p complexes.[31,103-106] In contrast, the Clns are not required for sporulation, and the destruction of Sic1p requires the sporulation-specific Ime2p protein kinase.[101] Ime2p is not the *CLB5/6* kinase that is required for DNA replication however, since a strain lacking Sic1p can complete DNA replication in the absence of Ime2p.[101] These results show that at least some mitotic Cdc28p functions can be performed by sporulation specific protein kinases during sporulation.

In contrast to premeiotic DNA synthesis, there is clear evidence that Cdc28p is required for the meiotic divisions. *cdc28-4* diploids sporulated at the restrictive temperature arrest in pachytene (just prior to chromosome segregation).[100] In addition there is good genetic evidence demonstrating a requirement for B-type cyclins in MI and II.[107] However, the cyclins appear to have specialized meiotic requirements. *CLB1* appears to be the major Clb that controls exit from pachytene and entry into meiosis I, *CLB1,CLB3,* and *CLB4* appear to play partially redundant roles in promoting meiosis II, and *CLB2,* while required for mitosis, does not appear to play any role in meiosis.[107]

The transition from pachytene into meiosis I represents a major point of regulatory control in sporulation and there is good evidence that much of this regulation is achieved by controlling Cdc28p activities. For example, signals that control the exit from pachytene and entry into meiosis I include the completion of genetic recombination, which is monitored by a pathway termed the pachytene checkpoint (also referred to as the recombination checkpoint). If recombination has not been completed, this checkpoint pathway blocks the program in pachytene (for review see 108). *SWE1,* which encodes the kinase that phosphorylates the inhibitory tyrosine in the ATP binding pocket of Cdc28p is required for the pachytene checkpoint and Swe1p is activated in cells that have undergone checkpoint-mediated arrest.[109] An overlapping control mechanism that regulates Cdc28p activity is the transcriptional program of sporulation.[110,111] This transcriptional cascade involves the induction of several hundred promoters that can be divided into "early", "middle" and "late" temporal classes. Early genes are expressed when premeiotic DNA synthesis, synaptonemal complex formation, and recombination are occurring. Middle genes are expressed as cells exit pachytene, enter the nuclear divisions and begin to form spores. Late genes are expressed as spore formation is being completed. Thus, the key events of sporulation are tightly coupled to and controlled by this genetically programmed transcriptional cascade. The *CLB1, CLB3-CLB6* genes are transcriptionally induced as middle genes. The sporulation-specific Ndt80 transcription factor is required for this induction.[112,113] Ndt80 is negatively regulated by the pachytene checkpoint.[112-114] In checkpoint-arrested cells, middle gene induction and the wave of CLB transcription does not occur as Ndt80 is inhibited. It is likely that the negative regulation of CLB transcriptional induction during meiosis plays a role in preventing the nuclear divisions in checkpoint arrested cells.

Mutants in *cak1* show a variety of meiotic defects. Temperature shift experiments show that *CAK1* is required for the meiotic nuclear divisions[66] (EW unpublished data). These results are similar to those seen with conditional *CDC28* mutants,[100] and are consistent with the notion that Cak1p activates Cdc28p during meiosis. These results imply that the pool of activated Cdc28p present in prophase is insufficient to drive the nuclear divisions and raise the possibility that Cak1p can be rate limiting for the nuclear divisions in wild-type cells. Experiments to monitor the phosphorylation state of Cdc28p during meiotic development have not been reported but might help to clarify this issue.

MI and MII in yeast are rapidly followed by spore formation. This morphogenetic program initiates as a thickening of the outer plaque of the centrosome (referred to as the spindle pole body in yeast). Vesicular fusion mediated through a sporulation-specific arm of the secretory pathway, is involved in the outgrowth of the prospore membrane,[115] which surrounds the haploid meiotic products. Subsequently, spores-pecific components are assembled from within and around the double-layered prospore membrane to generate the spore wall (reviewed in 99).

SMK1 encodes a sporulation-specific MAP kinase homolog that is a central regulator of spore morphogenesis.[116] *smk1* null mutants complete meiosis I and II but are defective in assembling spore walls. Moreover, different *smk1* hypomorphs block the program at distinct steps in the morphogenic pathway.[117] *CAK1* in high copy number suppresses the spore morphogenesis defects of weakened *smk1* mutants.[66] Furthermore a *cak1* strain that is able to complete meiosis I and II but that is specifically defective for spore morphogenesis has been isolated. In addition, *CAK1* is transcriptionally induced when meiosis and spore morphogenesis are occurring. These results indicate that Cak1p is required not only for the nuclear divisions (a Cdc28p-dependent function) but also plays a role in activating the *SMK1* spore morphogenesis pathway. More recently it has been shown that Cak1p and Smk1p interact using a two-hybrid system in mitotic cells.[65] However, it has not been determined whether this interaction is direct or whether the Cak1p and Smk1p are tethered by another protein.

The expression of Cak1p protein has been examined and shown to change dramatically during sporulation.[95] During the early phase of sporulation, when DNA replication is occurring, Cak1p levels are comparable to the levels seen in mitotic cells. Subsequently Cak1p levels fall to near background during late prophase (around pachytene). These data suggest that the level of Cak1p protein during sporulation is regulated not only at the transcriptional level but also by a developmentally regulated proteolysis pathway. Subsequently Cak1p levels increase as middle genes are transcriptionally induced, and as cells exit pachytene and enter the meiotic divisions. Since Cak1p is required for Cdc28p activity and Cdc28p is rate limiting for exit from pachytene this might suggest that the regulation of Cak1p levels could play a role in the pachytene checkpoint. However, expression of Cak1p using a high level promoter that generates constitutive *CAK1* mRNA and protein levels does not cause bypass of the pachytene checkpoint (EW unpublished data).

Additional insight into the role of Cak1p in sporulation comes from the analysis of genetic backgrounds in which the mitotic role of *CAK1* has been made dispensable.[82] These studies made use of *CDC28-43244*, a multiply mutant form of Cdc28p, which lacks a phosphorylatable residue at position 169 (Thr-169 is the in vivo phosphorylation target of Cak1p) and additional substitutions that presumably hyperactivate the kinase. *CDC28-43244* mutant cells grow in the complete absence of *CAK1*. While homozygous *CDC28-43244 CAK1* diploids sporulate, a *CDC28-43244 cak1-Δ* homozygous diploid is sporulation-defective (EW unpublished data). Since the Cdc28-43244 mutant protein lacks the threonine that is phosphorylated by Cak1p, these results imply that Cak1p functions in sporulation by phosphorylating a target other than Cdc28p. Biochemical studies indicate that Smk1p is inactive during sporulation of the *CDC28-43244 cak1-Δ* background suggesting that Cak1p activates a component of the *SMK1* MAPK pathway (M. Shaber and EW unpublished data). Kin28p, the other known mitotic target of Cak1p, does not appear to function in activating Smk1p since a *kin28* mutant lacking the residue normally phosphorylated by Cak1p in vegetative cells sporulates normally (J. Kimmelman, and M. Solomon personal communication).

In summary, Cak1p is highly regulated and plays multiple roles during sporulation. First it appears to be required for exit from pachytene and completion of the meiosis I and II divisions. This observation is consistent with the central role of Cdc28p in nuclear division. In addition, Cak1p appears to activate the *SMK1* pathway by a Cdc28p-independent mechanism. This later sporulation-specific function has led to the suggestion that Cak1p plays a role in coordinating meiosis with the spore differentiation pathway.[66]

Removal of the Activating Phosphorylation from Cdc28p

Compared to our knowledge of CAK, much less is known about the protein phosphatases that reverse the activating phosphorylation in CDKs. Previous studies in *S. pombe* and in *Xenopus* egg extracts raised the possibility that the dephosphorylation of this residue may be required for exit from mitosis,[118,119] and implicated type 2A and type 1 protein phosphatases in the dephosphorylation of Cdc2.[119,120] More recently, a dual specificity phosphatase, KAP (also called Cdi1, Cip2),[36,121,122] was shown to preferentially dephosphorylate monomeric CDK2 in vitro.[123] This result is consistent with the observation that *Xenopus* Cdc2 is dephosphorylated only after cyclin degradation.[119] However, there is no obvious KAP homolog in budding yeast. Using Thr-169 phosphorylated Cdc28p as substrate, we biochemically identified the Cdc28p phosphatase as belonging to the type 2C family of Ser/Thr protein phosphatases (PP2C): this activity required Mg^{2+} and was insensitive to PP1/PP2A and dual-specificity/tyrosine phosphatase inhibitors such as microcystin-LR, vanadate, and tungstate.[124] Two yeast PP2Cs, Ptc2p and Ptc3p, display Cdc28p phosphatase activity in vitro and in vivo, and account for ~90% of Cdc28p phosphatase activity in yeast extracts. Overproduction of Ptc2p or Ptc3p reduces the level of Thr-169 phosphorylation in Cdc28p in vivo and results in synthetic lethality in temperaturesensitive *cak1* strains at the permissive temperature. Furthermore, the dual disruption of *PTC2* and *PTC3* suppresses the growth defect of a *cak1* mutant at semi-permissive temperature. Since the phosphorylation of Cdc28p is the only essential function of Cak1p,[82] Ptc2p and Ptc3p are likely to be the physiological Cdc28p phosphatases in budding yeast. Like KAP, Ptc2p and Ptc3p prefer monomeric CDKs rather than cyclin-bound CDKs as substrates. PP2C-like activities are also responsible for >99% of CDK2 phosphatase activity in HeLa cell extracts.[124] We recently demonstrated that these CDK2 phosphatase activities belong to PP2Cα and PP2Cβ2, the closest homologs of yeast Ptc2p and Ptc3p.[125] Therefore, the ability of PP2Cs to reverse the activating phosphorylation of CDKs is evolutionarily conserved.

The balance between activating phosphorylation and dephosphorylation of CDKs appears to vary greatly between species. For example, studies from *S. pombe* and *Xenopus* show that the activating phosphorylation of Cdc2 is removed rapidly either during or at the end of mitosis.[118,119] In budding yeast, however, Cdc28p remains phosphorylated at Thr-169 throughout the cell cycle.[3,87,97] This difference is likely due to the different substrate specificities of the respective CAKs and to the relative activities of CAK and PP2C in the different species: (i) budding yeast Cak1p can phosphorylate monomeric CDKs and Cdc28p molecules that are dephosphorylated after cyclin degradation.[77] (ii) CDK phosphatase activity in HeLa cell extracts is much higher than in budding yeast extracts. In budding yeast, dephosphorylation of Cdc28p by Ptc2p and Ptc3p happens less frequently than phosphorylation by Cak1p as indicated by the fact that most monomeric Cdc28p molecules are phosphorylated. Using a phosphospecific antibody, we estimated that phosphorylation of Cdc28p by Cak1p is about 9-fold more frequent than dephosphorylation by Ptc2p and Ptc3p combined.[97] In contrast, PP2C-like activities in HeLa cell extract are about 10-fold higher than those in yeast extract[124] but no comparison has been done of yeast and human CAK activities. Therefore, Cak1p plays a dominant role in determining the phosphorylation of monomeric Cdc28p in budding yeast whereas PP2Cs might play the dominant role in human cells.

The identification of the CDK phosphatases also raised two questions. First, does the removal of the activating phosphorylation have any regulatory role during the cell cycle? Second, how are these PP2Cs regulated? In budding yeast, a regulatory role for the removal of the activating phosphorylation from Cdc28p during the normal cell cycle has not been detected. Three independent lines of experiments do not currently support such a role:

1. dual deletion of PTC2 and PTC3 has no apparent phenotype,[124]
2. the activating phosphorylation of Cdc28p is constant during the cell cycle, and

3. the Cdc28-43244p mutant, which mimics a permanent "phosphorylated" form of Cdc28p (T169E with several additional mutations), supports cell growth.[82]

However, it is unclear whether the same scenario occurs in higher eukaryotes. Studies in *S. pombe* and *Xenopus* egg extracts raised the possibility that dephosphorylation at the site of the activating phosphorylation may be required for exit from mitosis.[118,119] Considering that CAK activity does not vary in cycling *Xenopus* egg extracts,[16] it will be interesting to determine if the higher eukaryotic cdk phosphatase activity is regulated during the cell cycle. Previous studies showed that PP2Cα might be phosphorylated at its C-terminus[126] and PP2Cβ mRNA levels and activity are up-regulated during the 1 alpha,25-dehydroxyvitamin D3-induced monocytic differentiation of leukemic HL-60 cells.[127] Further studies will be necessary to determine whether the extent of activating phosphorylation of CDKs is modulated by the activity of PP2Cs. Like Ptc2p and Ptc3p in budding yeast, human PP2Cα and β isoforms contain potential sites for N-terminal myristoylation. It will be interesting to determine whether the CDK phosphatase is myristoylated in vivo and how the myristoylation regulates PP2C and its localization.

The Activation Pathway of Cdc28p

Because CDKs are regulated by several mechanisms, including cyclin binding, activating phosphorylation by CAKs, inhibitory phosphorylation by WEE1-like kinases, and association with inhibitory proteins, there are many possible pathways that can produce the active form of the CDK. For the budding yeast CDK, Cdc28p, the combinatorial problem is simplified somewhat because inhibitory phosphorylation does not play a major role in Cdc28p regulation during normal mitoses.[48,49] Phosphorylation of the inhibitory site, Tyr-19, is required instead for a checkpoint that monitors the coordination of the budding and nuclear division cycles.[128-130] It is not known whether Tyr-19 phosphorylation in response to the checkpoint signal is dependent on prior cyclin binding or activating phosphorylation of Cdc28p; however, in other organisms, inhibitory phosphorylation occurs only after cyclin binding.

The physiological role of the Cdc28p inhibitors, Far1p and Sic1p, is to down-regulate the active Cdc28p kinase,[29,31,32,131] suggesting that they are late players in the activation pathway, associating with the CAK-phosphorylated, cyclin-bound Cdc28p complex. Both inhibitors have indeed been shown through biochemical experiments to bind and inhibit Cdc28p/cyclin complexes.[29,31]

The best studied aspects of the Cdc28p activation pathway concern the relative order of cyclin binding and activating phosphorylation on Thr-169 by Cak1p. Experiments using a variety of methods to resolve Thr-169 phosphorylated from unphosphorylated Cdc28p, have concluded that essentially all of the Cdc28p in the cell is phosphorylated on Thr-169 throughout the cell cycle.[87,97,132] Cdc28p is found in three forms in the cell:

1. as a monomer [inactive],
2. as heterodimer with cyclins [active], and
3. as heterotrimer with cyclins and inhibitors [inactive].

Other potential forms would be a heterodimer with inhibitors [inactive] and a high molecular form in complex with chaperones like Cdc37p [inactive]. Interestingly, there is a 10-fold excess of Cdc28p over cyclin molecules in any given phase of the cell cycle. The monomeric pool of Cdc28p, which comprises 80-90% of the total, is as highly phosphorylated as the pool of active Cdc28p/cyclin complexes,[97,133] a surprising observation because monomeric CDKs in higher eukaryotes are not phosphorylated.[16,134] Cak1p phosphorylates monomeric CDKs much more efficiently than CDK/cyclin complexes, and both Clb and Cln cyclins associate more strongly with prephosphorylated Cdc28p than with the unphosphorylated form.[77,97] Taken together, these results suggest that the pool of phosphorylated monomer seen in vivo is likely to represent Cak1p phosphorylated monomers, and that Cdc28p is activated by subsequent binding of cyclin to phosphorylated Cdc28p. Under the conditions investigated, dephosphorylation of Thr-169 has not been shown to be a significant mechanism of Cdc28p regulation.[82,97,124] The Thr-169 phosphatases, Ptc2p and Ptc3p, act primarily on monomeric Cdc28p[124] and since there is a large pool of phosphorylated Cdc28p monomers a limiting effect can be observed only when Cak1p function is compromised [see also section on Removal of the Activating Phosphorylation from Cdc28p].[124]

Studies on the interplay of Cdc28p regulatory mechanisms indicate that Cdc28p monomers are phosphorylated on Thr-169 by Cak1p soon after their synthesis. As cyclins accumulate, they bind to and activate the phosphorylated monomers. Cdc28p activity is switched off either through the destruction of the cyclin subunit or by binding of inhibitors to the active complex. Cdc28p activity can be further modulated through inhibitory phosphorylation, probably of the Thr-169 phosphorylated, cyclin-bound form, in response to the budding checkpoint signal, or through dephosphorylation of Cdc28p monomers by the Thr-169 phosphatases.

Although the fundamental mechanisms of Cdc28p regulation are conserved in other eukaryotes, certain aspects of the CDK activation pathway appears not to be. CDC2 and CDK2 in mammals are phosphorylated by CAK only after binding to cyclin.[16,77,81,134,135] Budding yeast Cak1p and the CDK7/cyclin H family of CAKs that phosphorylate CDC2 and CDK2 have divergent biochemical properties that suit them to their respective activation pathways. For example, Cak1p prefers to phosphorylate monomeric CDKs whereas CDK7/cyclin H CAKs prefer to phosphorylate CDK/cyclin complexes.[77,81] Also, Cak1p is found primarily in the cytoplasm,[95] colocalizing with the large pool of monomeric Cdc28p,[98] whereas CDK7/cyclin H is found in the nucleus where it can phosphorylate CDKs which are directed to the nucleus upon cyclin binding.[4,136-138]

One possible explanation for the opposite order of activation and for the evolution of divergent CAKs, at least for the mitosis promoting factor [MPF], CDC2, could stem from the relative importance of inhibitory phosphorylation in CDK regulation in budding yeast versus other eukaryotes. While inhibitory phosphorylation of Cdc28p is dispensable during normal cell cycles, it is critically important for the correct timing of mitosis in other organisms (for review, see 7, 139). Removal of the inhibitory phosphorylations from CDC2 is the final step necessary for mitotic entry in *Xenopus.*[15] Because inhibitory phosphorylation cannot occur until after cyclin binding,[15] one way to ensure that CDC2 is not prematurely activated is for activating phosphorylation to also depend on prior cyclin binding, and ideally, on prior phosphorylation of the inhibitory sites.

A second possible explanation for the unusual Cdc28p activation pathway may be differences in the behavior of the nuclear membrane during mitosis. In mammalian cells, CDKs are known to shuttle in and out of the nucleus and nuclear envelope breakdown occurs at the beginning of mitosis (reviewed in 4, 140, 141). Budding yeast, like other fungi carry out a closed mitosis in which the nuclear envelope does not break down. However, CAK-phosphorylated CDK/cyclin complexes are found both in the nucleus and the cytoplasm. To achieve this, without mixing their cytoplasmic and nuclear contents, budding yeast must either have CAK activity in both compartments, or as appears to be the case, phosphorylate their CDKs as monomers in the cytoplasm before they enter the nucleus. Intriguingly, *S. pombe*, which also has a closed mitosis, appears to have two CAKs—a CDK7/cyclin H-type CAK (Mcs6/Mcs2) and a Cak1p-type CAK (Csk1), although the physiological roles of these two enzymes have not yet been resolved.[73,75,76,142]

Future experiments aimed at understanding why different CDK activation pathways developed and continue to persist will give us insight into the complex and subtle variety of strategies of growth regulation used by eukaryotes. Results from such studies might explain the difference between CDK7-like and Cak1p-like CAKs.

Acknowledgments

We thank Mark Solomon for support, comments on the manuscript, and communication of unpublished results. Mike Shaber and Jonathan Kimmelman are acknowledged for communicating unpublished results.

P. K. thanks Karen Vousden and the NCI for support. A.C. is a Leukemia and Lymphoma society fellow.

References

1. King RW, Deshaies RJ, Peters J-M, Kirschner MW. How proteolysis drives the cell cycle. Science 1996; 274:1652-1659.
2. Morgan DO. The dynamics of cyclin dependent kinase structure. Curr Opin Cell Biol 1996; 8:767-772.
3. Morgan DO. Cyclin-dependent kinases: Engines, clocks, and microprocessors. Annu Rev Cell Dev Biol 1997; 13:261-291.
4. Pines J. Cyclins and cyclin-dependent kinases: A biochemical view. Biochem J 1995; 308:697-711.
5. Sherr CJ, Roberts JM. Inhibitors of mammalian G_1 cyclin-dependent kinases. Genes Dev 1995; 9:1149-1163.
6. Sherr CJ, Roberts JM. CDK inhibitors: Positive and negative regulators of G_1-phase progression. Genes Dev 1999; 13:1501-1512.
7. Solomon MJ, Kaldis P. Regulation of cdks by phosphorylation. In: Pagano M, ed. Results and Problems in Cell Differentiation "Cell Cycle Control". Heidelberg: Springer, 1998:79-109.
8. Peeper DS, Parker LL, Ewen ME, Toebes M, Hall FL, Xu M et al. A- and B-type cyclins differentially modulate substrate specificity of cyclin-cdk complexes. EMBO J 1993; 12:1947-1954.
9. Schulman BA, Lindstrom DL, Harlow E. Substrate recruitment to cyclin-dependent kinase 2 by a multipurpose docking site on cyclin A. Proc Natl Acad Sci USA 1998; 95:10453-10458.
10. Cross FR, Jacobson MD. Conservation and function of a potential substrate-binding domain in the yeast Clb5 B-type cyclin. Mol Cell Biol 2000; 20:4782-4790.
11. Diehl JA, Zindy F, Sherr CJ. Inhibition of cyclin D1 phosphorylation on threonine-286 prevents its rapid degradation via the ubiquitin-proteasome pathway. Genes Dev 1997; 11:957-972.
12. Draviam VM, Orrechia S, Lowe M, Pardi R, Pines J. The localization of human cyclins B1 and B2 determines CDK1 substrate specificity and neither enzyme requires MEK to disassemble the Golgi apparatus. J Cell Biol 2001; 152:945-958.
13. Pines J, Hunter T. The differential localization of human cyclins A and B is due to a cytoplasmic retention signal in cyclin B. EMBO J 1994; 13:3772-3781.
14. Pines J, Hunter T. Human cyclins A and B1 are differentially located in the cell and undergo cell cycle-depenent nuclear transport. J Cell Biol 1991; 115:1-17.
15. Solomon MJ, Glotzer M, Lee TH, Philippe M, Kirschner MW. Cyclin activation of $p34^{cdc2}$. Cell 1990; 63:1013-1024.
16. Solomon MJ, Lee T, Kirschner MW. Role of phosphorylation in $p34^{cdc2}$ activation: Identification of an activating kinase. Mol Biol Cell 1992; 3:13-27.
17. Evans T, Rosenthal ET, Youngblom J, Distel D, Hunt T. Cyclin: A protein specified by maternal mRNA in sea urchin eggs that is destroyed at each cleavage division. Cell 1983; 33:389-396.
18. Sherr CJ. G1 phase progression: Cycling on cue. Cell 1994; 79:551-555.
19. Koch C, Nasmyth K. Cell cycle regulated transcription in yeast. Curr Opin Cell Biol 1994; 6:451-459.
20. Patton EE, Willems AR, Tyers M. Combinatorial control in the ubiquitin-dependent proteolysis: Don't Skp the F-box hypothesis. Trends Gene 1998; 14:236-243.
21. Townsley FM, Ruderman JV. Proteolytic ratchets that control progression through mitosis. Trends Cell Biol 1998; 8:238-244.

22. Tyers M, Tokiwa G, Futcher B. Comparison of the *Saccharomyces cerevisiae* G_1 cyclins: Cln3 may be an upstream activator of Cln1, Cln2 and other cyclins. EMBO J 1993; 12:1955-1968.
23. Wittenberg C, Sugimoto K, Reed SI. G1-specific cyclins of *S. cerevisiae*: Cell cycle periodicity, regulation by mating pheromone, and association with the $p34^{CDC28}$ protein kinase. Cell 1990; 62:225-237.
24. Oehlen LJ, Cross FR. G1 cyclins CLN1 and CLN2 repress the mating factor response pathway at Start in the yeast cell cycle. Genes Dev 1994; 8:1058-1070.
25. Oehlen LJ, Jeoung DI, Cross FR. Cyclin-specific START events and the G1-phase specificity of arrest by mating factor in budding yeast. Mol Gen Genet 1998; 258:183-198.
26. Mendenhall MD, Hodge AE. Regulation of Cdc28 cyclin-dependent protein kinase activity during the cell cycle of the yeast *Saccharomyces cerevisiae*. Microbiol. Mol Biol Rev 1998; 62:1191-1243.
27. Chang F, Herskowitz I. Identification of a gene necessary for cell cycle arrest by a negative growth factor of yeast: FAR1 is an inhibitor of a G1 cyclin, CLN2. Cell 1990; 63:999-1011.
28. Jeoung DI, Oehlen LJ, Cross FR. Cln3-associated kinase activity in *Saccharomyces cerevisiae* is regulated by the mating factor pathway. Mol Cell Biol 1998; 18:433-441.
29. Peter M, Herskowitz I. Direct inhibition of the yeast cyclin-dependent kinase Cdc28-Cln by Far1. Science 1994; 265:1228-1231.
30. Mendenhall MD. An inhibitor of $p34^{CDC28}$ protein kinase activity from *Saccharomyces cerevisiae*. Science 1993; 259:216-219.
31. Schwob E, Böhm T, Mendenhall MD, Nasmyth K. The B-type cyclin kinase inhibitor $p40^{SIC1}$ controls the G1 to S transition in *S. cerevisiae*. Cell 1994; 79:233-244.
32. Donovan JD, Toyn JH, Johnson AL, Johnston LH. $p40^{SDB25}$, a putative CDK inhibitor, has a role in the M/G1 transition in *Saccharomyces cerevisiae*. Genes Dev 1994; 8:1640-1653.
33. Toyn JH, Johnson AL, Donovan JD, Toone WM, Johnston LH. The Swi5 transcription factor of *Saccharomyces cerevisiae* has a role in exit from mitosis through induction of the cdk-inhibitor Sic1 in telophase. Genetics 1997; 145:85-96.
34. El-Deiry WS, Tokino T, Velculescu VE, Levy DB, Parsons R, Trent JM et al. WAF1, a potent mediator of p53 tumor suppression. Cell 1993; 75:817-825.
35. Gu Y, Turck CW, Morgan DO. Inhibition of CDK2 activity in vivo by an associated 20K regulatory subunit. Nature 1993; 366:707-710.
36. Harper JW, Adami GR, Wei N, Keyomarski K, Elledge SJ. The p21 cdk-interacting protein Cip1 is a potent inhibitor of G1 cyclin-dependent kinases. Cell 1993; 75:805-816.
37. Polyak K, Kato J-y, Solomon MJ, Sherr CJ, Massagué J, Roberts JM et al. $p27^{Kip1}$, a cyclin-cdk inhibitor, links transforming growth factor-β and contact inhibition to cell cycle arrest. Genes Dev 1994; 8:9-22.
38. Polyak K, Lee M-H, Erdjument-Bromage H, Koff A, Roberts JM, Tempst P, et al. Cloning of $p27^{Kip1}$, a cyclin-dependent kinase inhibitor and a potential mediator of extracellular antimitogenic signals. Cell 1994; 78:59-66.
39. Toyoshima H, Hunter T. p27, a novel inhibitor of G1 cyclin-cdk protein kinase activity, is related to p21. Cell 1994; 78:67-74.
40. Lee MH, Reynisdóttir I, Massagué J. Cloning of $p57^{KIP2}$, a cyclin-dependent kinase inhibitor with unique domain structure and tissue distribution. Genes Dev 1995; 9:639-649.
41. Matsuoka S, Edwards M, Bai C, Parker S, Zhang P, Baldini A et al. $p57^{KIP2}$, a structurally distinct member of the $p21^{CIP1}$ cdk inhibitor family, is a candidate tumor suppressor gene. Genes Dev 1995; 9:650-662.
42. Hannon GJ, Beach D. $p15^{INK4B}$ is a potential effector of TGF-b-induced cell cycle arrest. Nature 1994; 371:257-261.
43. Serrano M, Hannon GJ, Beach D. A new regulatory motif in cell cycle control causing specific inhibition of cyclin D/CDK4. Nature 1993; 366:704-707.
44. Guan K, Jenkins CW, Li Y, Nichols MA, Wu X, O'Keefe CL et al. Growth suppression by p18, A $p16^{INK4/MTS1}$- and $p14^{INK4B/MTS2}$- related CDK6 inhibitor, correlates with wild-type pRb function. Genes Dev 1994; 8:2939-2952.

45. Chan FKM, Zhang J, Chen L, Shapiro DN, Winoto A. Identification of human and mouse p19, a novel CDK4 and CDK6 inhibitor with homology to p16^{ink4}. Mol Cell Biol 1995; 15:2682-2688.
46. Hirai H, Roussel MF, Kato J-y, Ashmun RA, Sherr CJ. Novel INK4 proteins, p19 and p18, are specific inhibitors of the cyclin D-dependent kinases CDK4 and CDK6. Mol Cell Biol 1995; 15:2672-2681.
47. Coleman TR, Dunphy WG. Cdc2 regulatory factors. Curr Opin Cell Biol 1994; 6:877-882.
48. Amon A, Surana U, Muroff I, Nasmyth K. Regulation of p34^{CDC28} tyrosine phosphorylation is not required for entry into mitosis in *S. cerevisiae*. Nature 1992; 355:368-371.
49. Sorger PK, Murray AW. S-phase feedback control in budding yeast independent of tyrosine phosphorylation of p34^{CDC28}. Nature 1992; 355:365-368.
50. Lew DJ. Cell-cycle checkpoints that ensure coordination between nuclear and cytoplasmic events in *Saccharomyces cerevisiae*. Curr Opin Gen Dev 2000; 10:47-53.
51. Johnson LN, Noble MEM, Owen DJ. Active and inactive protein kinases: Structural basis for regulation. Cell 1996; 85:149-158.
52. Espinoza FH, Farrell A, Erdjument-Bromage H, Tempst P, Morgan DO. A cyclin-dependent kinase-activating kinase (CAK) in budding yeast unrelated to vertebrate CAK. Science 1996; 273:1714-1717.
53. Fesquet D, Labbé J-C, Derancourt J, Capony J-P, Galas S, Girard F et al. The *MO15* gene encodes the catalytic subunit of a protein kinase that activates cdc2 and other cyclin-dependent kinases (CDKs) through phosphorylation of Thr161 and its homologues. EMBO J 1993; 12:3111-3121.
54. Kaldis P, Sutton A, Solomon MJ. The CDK-activating kinase (CAK) from budding yeast. Cell 1996; 86:553-564.
55. Poon RYC, Yamashita K, Adamczewski JP, Hunt T, Shuttleworth J. The cdc2-related protein p40^{MO15} is the catalytic subunit of a protein kinase that can activate p33^{cdk2} and p34^{cdc2}. EMBO J 1993; 12:3123-3132.
56. Solomon MJ, Harper JW, Shuttleworth J. CAK, the p34^{cdc2} activating kinase, contains a protein identical or closely related to p40^{MO15}. EMBO J 1993; 12:3133-3142.
57. Thuret J-Y, Valay J-G, Faye G, Mann C. Civ1 (CAK in vivo), a novel CDK-activating kinase. Cell 1996; 86:565-576.
58. Kaldis P. The CDK-activating kinase: From yeast to mammals. Cell Mol Life Sci 1999; 55:284-296.
59. Elledge SJ, Spottswood MR. A new human p34 protein kinase, CDK2, identified by complementation of a *cdc28* mutation in *Saccharomyces cerevisiae*, is a homolog of *Xenopus* Eg1. EMBO J 1991; 10:2653-2659.
60. Simon M, Seraphin B, Faye G. Kin28, a yeast split gene coding for a putative protein kinase homologous to CDC28. EMBO J 1986; 5:2697-2701.
61. Valay JG, Simon M, Faye G. The Kin28 protein kinase is associated with a cyclin in *Saccharomyces cerevisiae*. J Mol Biol 1993; 234:307-310.
62. Cismowski MJ, Laff GM, Solomon MJ, Reed SI. *KIN28* encodes a C-terminal domain kinase that controls mRNA transcription in *Saccharomyces cerevisiae* but lacks cyclin-dependent kinase-activating kinase (CAK) activity. Mol Cell Biol 1995; 15:2983-2992.
63. Valay J-G, Simon M, Dubois M-F, Bensaude O, Facca C, Faye G. The *KIN28* gene is required both for RNA polymerase II mediated transcription and phosphorylation of the Rpb1p CTD. J Mol Biol 1995; 249:535-544.
64. Feaver WJ, Svejstrup JQ, Henry NL, Kornberg RD. Relationship of CDK-activating kinase and RNA polymerase II CTD kinase TFIIH/TFIIK. Cell 1994; 79:1103-1109.
65. Uetz P, Giot L, Cagney G, Mansfield TA, Judson RS, Knight JR et al. A comprehensive analysis of protein-protein interactions in *Saccharomyces cerevisiae*. Nature 2000; 403:623-627.
66. Wagner M, Pierce M, Winter E. The CDK-activating kinase *CAK1* can dosage suppress sporulation defects of *smk1* MAP kinase mutants and is required for spore wall morphogenesis in *Saccharomyces cerevisiae*. EMBO J 1997; 16:1305-1317.
67. Sutton A, Freiman R. The Cak1p protein kinase is required at G_1/S and G_2/M in the budding yeast cell cycle. Genetics 1997; 147:57-71.

68. Valay JG. Etude génétique et physiologique de la protéine kinase Kin28p de *Saccharomyces cerevisiae*. Paris: University of Paris VI, 1994.
69. Chun KT, Goebl MG. Mutational analysis of Cak1p, an essential protein kinase that regulates cell cycle progression. Mol Gen Genet 1997; 256:365-375.
70. Fernandez-Sarabia MJ, Sutton A, Zhong T, Arndt KT. SIT4 protein phosphatase is required for the normal accumulation of *SWI4*, *CLN1*, *CLN2* and *HCS26* RNAs during late G1. Genes Dev 1992; 6:2417-2428.
71. Sutton A, Immanuel D, Arndt KT. The Sit4 protein phosphatase functions in late G1 for progression into S phase. Mol Biol Cell 1991; 11:2133-2148.
72. Faye G, Valay JG, Mann C, Thuret J-Y. patent application WO 99/07836, PCT/FR 98/01788. 1999.
73. Hermand D, Pihlak A, Westerling T, Damagnez V, Vanderhaute J, Cottarel G et al. Fission yeast Csk1 is a CAK-activating kinase (CAKAK). EMBO J 1998; 17:7230-7238.
74. Umeda M, Bhalerao RP, Schell J, Uchimiya H, Koncz C. A distinct cyclin-dependent kinase-activating kinase of *Arabidopsis thaliana*. Proc Natl Acad Sci USA 1998; 95:5021-5026.
75. Lee KM, Saiz JE, Barton WA, Fisher RP. Cdc2 activation in fission yeast depends on Mcs6 and Csk1, two partially redundant CDK-activating kinases (CAKs). Curr Biol 1999; 9:441-444.
76. Hermand D, Westerling T, Pihlak A, Thuret J-Y, Vallenius T, Tiainen M et al. Specificity of cdk activation *in vivo* by the two CAKs Mcs6 and Csk1 in fission yeast. EMBO J 2001; 20:82-90.
77. Kaldis P, Russo AA, Chou HS, Pavletich NP, Solomon MJ. Human and yeast CDK-activating kinases (CAKs) display distinct substrate specificities. Mol Biol Cell 1998; 9:2545-2560.
78. Egan EA, Solomon MJ. Cyclin-stimulated binding of Cks proteins to cyclin-dependent kinases. Mol Cell Biol 1998; 18:3659-3667.
79. Kaldis P, Cheng A, Solomon MJ. The effects of changing the site of activating phosphorylation in cdk2 from threonine to serine. J Biol Chem 2000; 275:32578-32584.
80. Hagopian JC, Kirtley MP, Stevenson LM, Gergis RM, Russo AA, Pavletich NP et al. Kinetic basis for activation of cdk2/cyclin A by phosphorylation. J Biol Chem 2001; 276:275-280.
81. Fisher RP, Morgan DO. A novel cyclin associates with MO15/CDK7 to form the CDK-activating kinase. Cell 1994; 78:713-724.
82. Cross FR, Levine K. Molecular evolution allows bypass of the requirement for activation loop phosphorylation of the Cdc28 cyclin-dependent kinase. Mol Cell Biol 1998; 18:2923-2931.
83. Adamczewski JP, Rossignol M, Tassan J-P, Nigg EA, Moncollin V, Egly J-M. MAT1, cdk7 and cyclin H form a kinase complex which is UV light-sensitive upon association with TFIIH. EMBO J 1996; 15:1877-1884.
84. Roy R, Adamczewski JP, Seroz T, Vermeulen W, Tassan J-P, Schaeffer L et al. The MO15 cell cycle kinase is associated with the TFIIH transcription-DNA repair factor. Cell 1994; 79:1093-1101.
85. Serizawa H, Mäkelä TP, Conaway JW, Conaway RC, Weinberg RA, Young RA. Association of CDK-activating kinase subunits with transcription factor TFIIH. Nature 1995; 374:280-282.
86. Shiekhattar R, Mermelstein F, Fisher RP, Drapkin R, Dynlacht B, Wessling HC et al. CDK-activating kinase complex is a component of human transcription factor TFIIH. Nature 1995; 374:283-287.
87. Espinoza FH, Farrell A, Nourse JL, Chamberlin HM, Gileadi O, Morgan DO. Cak1 is required for Kin28 phosphorylation and activation in vivo. Mol Cell Biol 1998; 18:6365-6373.
88. Kimmelman J, Kaldis P, Hengartner CJ, Laff GM, Koh SS, Young RA et al. Activating phosphorylation of the Kin28p subunit of yeast TFIIH by Cak1p. Mol Cell Biol 1999; 19:4774-4787.
89. Faye G, Simon M, Valay JG, Fesquet D, Facca C. Rig2, a RING finger protein that interacts with the Kin28/Ccl1 CTD kinase in yeast. Mol Gen Genet 1997; 255:460-466.
90. Feaver WJ, Henry NL, Wang Z, Wu X, Svejstrup JQ, Bushnell DA et al. Genes for Tfb2, Tfb3, and Tfb4 subunits of yeast transcription/repair factor IIH. J Biol Chem 1997; 272:19319-19327.

91. Hemmer W, McGlone M, Tsigelny I, Taylor SS. Role of the glycine triad in the ATP-binding site of cAMP-dependent protein kinase. J Biol Chem 1997; 272:16946-16954.
92. Enke DA, Kaldis P, Holmes JK, Solomon MJ. The CDK-activating kinase (Cak1p) from budding yeast has an unusual ATP-binding pocket. J Biol Chem 1999; 274:1949-1956.
93. Gibbs CS, Zoller MJ. Rational scanning mutagenesis of a protein kinase identifies functional regions involved in catalysis and substrate interaction. J Biol Chem 1991; 266:8923-8931.
94. Enke DA, Kaldis P, Solomon MJ. Kinetic analysis of the cyclin-dependent kinase-activating kinase (Cak1p) from budding yeast. J Biol Chem 2000; 275:33267-33271.
95. Kaldis P, Pitluk ZW, Bany IA, Enke DA, Wagner M, Winter E et al. Localization and regulation of the CDK-activating kinase (Cak1p) from budding yeast. J Cell Sci 1998; 111:3585-3596.96.
96. Russo AA, Jeffrey PD, Patten AK, Massagué J, Pavletich NP. Crystal structure of the $p27^{Kip1}$ cyclin-dependent kinase inhibitor bound to the cyclin A-Cdk2 complex. Nature 1996; 382:325-331.
97. Ross KE, Kaldis P, Solomon MJ. Activating phosphorylation of the *Saccharomyces cerevisiae* cyclin-dependent kinase, Cdc28p, precedes cyclin binding. Mol Biol Cell 2000; 11:1597-1609.
98. Wittenberg C, Richardson SL, Reed SI. Subcellular localization of a protein kinase required for cell cycle initiation in *Saccharomyces cerevisiae*: Evidence for an association between the CDC28 gene product and the insoluble cytoplasmic matrix. J Cell Biol 1987; 105:1527-1538.
99. Kupiec M, Byers B, Esposito R, Mitchell A. Meiosis and sporulation in *Saccharomyces cerevisiae*. In: Pringle J, Broach J, Jones E, eds. The Molecular and Cellular Biology of the Yeast *Saccharomyces*. Cold Spring Harbor: Cold Spring Harbor Press, 1997:889-1036.
100. Shuster EO, Byers B. Pachytene arrest and other meiotic effects of the start mutations in *Saccharomyces cerevisiae*. Genetics 1989; 123:29-43.
101. Dirick L, Goetsch L, Ammerer G, Byers B. Regulation of meiotic S phase by Ime2 and a Clb5,6-associated kinase in *Saccharomyces cerevisiae*. Science 1998; 281:1854-1857.
102. Stuart D, Wittenberg C. CLB5 and CLB6 are required for premeiotic DNA replication and activation of the meiotic S/M checkpoint. Genes Dev 1998; 12:2698-2710.
103. Dirick L, Böhm T, Nasmyth K. Roles and regulation of Cln-Cdc28 kinases at the start of the cell cycle of *Saccharomyces cerevisiae*. EMBO J 1995; 14:4803-4813.
104. Schneider BL, Yang Q-H, Futcher AB. Linkage of replication to start by the cdk inhibitor Sic1. Science 1996; 272:560-562.
105. Tyers M. The cyclin-dependent kinase inhibitor $p40^{SIC1}$ imposes the requirement for Cln G1 cyclin function at Start. Proc Natl Acad Sci USA 1996; 93:7772-7776.
106. Verma R, Annan RS, Huddleston MJ, Carr SA, Reynard G, Deshaies RJ. Phosphorylation of Sic1p by G_1 cdk required for its degradation and entry into S phase. Science 1997; 278:455-460.
107. Dahmann C, Futcher B. Specialization of B-type cyclins for mitosis or meiosis in *S. cerevisiae*. Genetics 1995; 140:957-963.
108. Roeder GS, Bailis JM. The pachytene checkpoint. Trends Genet 2000; 16:395-403.
109. Leu JY, Roeder GS. The pachytene checkpoint in *S. cerevisiae* depends on Swe1-mediated phosphorylation of the cyclin-dependent kinase Cdc28. Mol Cell 1999; 4:805-814.
110. Chu S, DeRisi J, Eisen M, Mulholland J, Botstein D, Brown PO et al. The transcriptional program of sporulation in budding yeast [published erratum appears in Science 1998 Nov 20; 282(5393):1421]. Science 1998; 282:699-705.
111. Primig M, Williams RM, Winzeler EA, Tevzadze GG, Conway AR, Hwang SY et al. The core meiotic transcriptome in budding yeasts. Nat Genet 2000; 26:415-423.
112. Chu S, Herskowitz I. Gametogenesis in yeast is reglated by a transcriptional cascade dependent on Ndt80. Mol Cell 1998; 1:685-696.
113. Hepworth SR, Friesen H, Segall J. *NDT80* and the meiotic recombination checkpoint regulate expression of middle sporulation-specific genes in *Saccharomyces cerevisiae*. Mol Cell Biol 1998; 18:5750-5761.
114. Tung KS, Hong EJ, Roeder GS. The pachytene checkpoint prevents accumulation and phosphorylation of the meiosis-specific transcription factor Ndt80. Proc Natl Acad Sci USA 2000; 97:12187-12192.

115. Neiman AM, Stevenson BJ, Xu HP, Sprague GF, Jr., Herskowitz I, Wigler M et al. Functional homology of protein kinases required for sexual differentiation in *Schizosaccharomyces pombe* and *Saccharomyces cerevisiae* suggests a conserved signal transduction module in eukaryotic organisms. Mol Biol Cell 1993; 4:107-120.
116. Krisak L, Strich R, Winters RS, Hall JP, Mallory MJ, Kreitzer D et al. *SMK1*, a developmentally regulated MAP kinase, is required for spore wall assembly in *Saccharomyces cerevisiae*. Genes Dev 1994; 8:2151-2161.
117. Wagner M, Briza P, Pierce M, Winter E. Distinct steps in yeast spore morphogenesis require distinct SMK1 MAP kinase thresholds. Genetics 1999; 151:1327-1340.
118. Gould KL, Moreno S, Owen DJ, Sazer S, Nurse P. Phosphorylation at Thr167 is required for *Schizosaccharomyces pombe* $p34^{cdc2}$ function. EMBO J 1991; 10:3297-3309.
119. Lorca T, Labbé J-C, Devault A, Fesquet D, Capony J-P, Cavadore J-C et al. Dephosphorylation of cdc2 on threonine 161 is required for cdc2 kinase inactivation and normal anaphase. EMBO J 1992; 11:2381-2390.
120. Lee TH, Solomon MJ, Mumby MC, Kirschner MW. INH, a negative regulator of MPF, is a form of protein phosphatase 2A. Cell 1991; 64:415-423.
121. Gyuris J, Golemis E, Chertkov H, Brent R. Cdi1, a human G1 and S phase protein phosphatase that associates with Cdk2. Cell 1993; 75:791-803.
122. Hannon GJ, Casso D, Beach D. KAP: A dual specificity phosphatase that interacts with cyclin- dependent kinases. Proc Natl Acad Sci USA 1994; 91:1731-1735.
123. Poon YCR, Hunter T. Dephosphorylation of Cdk2 Thr^{160} by the cyclin-dependent kinase-interacting phosphatase KAP in the absence of cyclin. Science 1995; 270:90-93.
124. Cheng A, Ross KE, Kaldis P, Solomon MJ. Dephosphorylation of cyclin-dependent kinases by type 2C protein phosphatases. Genes Dev 1999; 13:2946-2957.
125. Cheng A, Kaldis P, Solomon MJ. Dephosphorylation of human cyclin-dependent kinases by protein phosphatase type 2Cα and β2 isoforms. J Biol Chem 2000; 275:34744-34749.
126. Kobayashi T, Kusuda K, Ohnishi M, Wang H, Ikeda S, Hanada M et al. Isoform specific phosphorylation of protein phosphatase 2C expressed in COS7 cells. FEBS Lett 1998; 430:222-226.
127. Nishikawa M, Omay SB, Nakai K, Kihira H, Kobayashi T, Tamura S et al. Up-regulation of protein serine/threonine phosphatase type 2C during 1 alpha,25-dihydroxyvitamin D3-induced monocytic differentiation of leukemic HL-60 cells. FEBS Lett 1995; 375:299-303.
128. Barral Y, Parra M, Bidlingmaier S, Snyder M. Nim1-related kinases coordinate cell cycle progression with the organization of the peripheral cytoskeleton in yeast. Genes Dev 1999; 13:176-187.
129. Lew DJ, Reed SI. A cell cycle checkpoint monitors cell morphology in budding yeast. J Cell Biol 1995; 129:739-749.
130. Sia RA, Herald HA, Lew DJ. Cdc28 tyrosine phosphorylaion and the morphogenesis checkpoint in budding yeast. Mol Biol Cell 1996; 7:1657-1666.
131. Peter M, Gartner A, Horecka J, Ammererer G, Herskowitz I. FR1 links the signal transduction pathway to the cell cycle machinery in yeast. Cell 1993; 73:747-760.
132. Hadwiger JA, Reed SI. Invariant phosphorylation of the *Saccharomyces cerevisiae* Cdc28 protein kinase. Mol Cell Biol 1988; 8:2976-2979.
133. Wittenberg C, Reed SI. Control of the yeast cell cycle is associated with assembly/disassembly of the Cdc28 protein kinase complex. Cell 1988; 54:1061-1072.
134. Desai D, Gu Y, Morgan DO. Activation of human cyclin-dependent kinases in vitro. Mol Biol Cell 1992; 3:571-582.
135. Desai D, Wessling HC, Fisher RP, Morgan DO. The effect of phosphorylation by CAK on cyclin binding by CDC2 and CDK2. Mol. Cell. Biol. 1995; 15:345-350.
136. Jordan P, Cunha C, Carmo-Fonseca M. The cdk7-cyclin H-MAT1 complex associated with TFIIH is localized in coiled bodies. Mol Biol Cell 1997; 8:1207-1217.
137. Tassan J-P, Schultz SJ, Bartek J, Nigg EA. Cell cycle analysis of the activity, subcellular localization, and subunit composition of human CAK (CDK-activating kinase). J Cell Biol 1994; 127:467-478.

138. Tassan J-P, Jaquenoud M, Fry AM, Frutiger S, Hughes GJ, Nigg EA. *In vitro* assembly of a functional human CDK7-cyclin H complex requires MAT1, a novel 36 kDa RING finger protein. EMBO J 1995; 14:5608-5617.
139. Lew DJ, Kornbluth S. Regulatory roles of cyclin dependent kinase phosphorylation in cell cycle control. Curr Opin Cell Biol 1996; 8:795-804.
140. Pines J. Four-dimensional control of the cell cycle. Nat Cell Biol 1999; 1:E73-E79.
141. Yang J, Kornbluth S. All aboard the cyclin train: Sucellular trafficking of cyclins and their cdk partners. Trends Cell Biol 1999; 9:207-210.
142. Damagnez V, Mäkelä TP, Cottarel G. *Schizosaccharomyces pombe* Mop1-Mcs2 is related to mammalian CAK. EMBO J 1995; 14:6164-6172.

Chapter 3

CDK-Activation in Fission Yeast *Schizosaccharomyces pombe*: Specificity Mediated by Distinct CAK Kinases

Thomas Westerling, Damien Hermand and Tomi P. Mäkelä

Introduction

Activation of cyclin dependent kinases (CDKs) by phosphorylation of the conserved T-loop is a regulatory step of cell cycle progression that was discovered just a decade ago. A wealth of data regarding this activation and of potential CDK activating kinases (CAKs) has accumulated. Despite this several crucial unresolved issues remain, including the unambiguous identification of the in vivo CAK kinase(s) in mammalian species. Here we focus on recent advances in this field in fission yeast *Schizosaccharomyces pombe* and discuss the implications of these results in a broader context.

Characteristics of the Fission Yeast Cell Cycle

The cell cycle of the fission yeast *S. pombe* is apparently significantly less complicated than that in higher organisms and also different from the other common yeast model organism, budding yeast *Saccharomyces cerevisiae*. In *S. pombe* one major CDK, namely Cdc2, is the kinase driving the cell cycle.[1] Furthermore the Cdc13 B-type cyclin normally associated with Cdc2 at mitosis can apparently drive the entire *S. pombe* cell cycle in a strain deleted for other reported B-type cyclins.[2] The minimal requirement of the Cdc13 cyclin leads to a model where the level of CDK activity rather than the specificity of the CDK-cyclin complexes is critical for cell cycle progression.[1] This might reflect a primitive cell cycle of a hypothetical ancestor eukaryote where replication and mitosis might have been concomitant due to a small genome size.[3] Increase of the DNA content leads to a separation of S-phase and mitosis dependent on the level of Cdc2 kinase activity, S-phase being initiated with a lower level than mitosis.[1] Keeping these differences in mind might give us clues to the reason why various organisms differ in fundamental aspects of CDK regulation such as CDK activation by phosphorylation.

The minimal requirement of the Cdc13 cyclin may also in part reflect the exotic life cycle of *S. pombe*. Although *S. pombe* goes through a diploid life cycle, it spends most of its time in haploid form not only in the laboratory but also in nature, e.g., in grapes and molasses.[4-6] Perhaps due to this habit *S. pombe* spends most (ca. 70%) if its division cycle in G2[7] as opposed to other model organisms that are predominantly G1 dwellers spending most of their time in the cell cycle in preparation of passing START.[8] A potential advantage of the G2 dominated cell cycle of the haploid *S. pombe* is the increased time spent with a 2C (G2) DNA content that

The CDK-Activating Kinase (CAK), edited by Philipp Kaldis.

is less prone to deleterious DNA alterations. Another consequence of this cell cycle distribution that is important to bear in mind when interpreting cell cycle experiments in *S. pombe* is the fact that at the time of visible daughter cell division (through septation) cells have already passed G1 in exponential cultures. This is because the size requirement of initiating S-phase is already achieved by the time mitosis is completed.[9-11]

In spite of these differences *S. pombe* has a typical eukaryote cell cycle with ordered progression through G1, S, G2 and M-phases in contrast to *S. cerevisiae* that does not appear to have a clear temporal separation of S-phase and M-phase events.[12] In addition, despite the apparent totipotency of Cdc2-Cdc13 in mutant *S. pombe* strains, other cyclins besides Cdc13 are important in normal G1-S regulation, although they are non-essential. Thus Cig2[13] is the major partner of Cdc2 in the G1 phase and regulates G1-S transition.[14] Also Cigl and Pucl cyclins contribute to G1 progression.[2,15] The recently characterized non-essential Rem1 cyclin may be more involved in meiotic events (José Ayté and Paul Nurse, personal communication).

Regulation of *S. pombe* Cdc2

The *S. pombe* mode of growth through bipolar apical extension and the resulting pen-shaped morphology[16] became the basis of classic genetic screens for genes involved in cell division cycle control. *S. pombe* cells unable to pass the G2/M checkpoint continue growth in the absence of division and thus increase cell length. On the other hand, if a mutation ablates the cell size checkpoint at G2/M the result is premature commitment to division with a cell size that is less than wild type. By employing these mutant phenotypes, a number of cell division cycle (cdc)[9] and *wee1*[17-19] mutants were isolated. The mutants include various components crucial for metabolic events involved in proliferation and a series of molecules associated with direct control of cell cycle progression (for review see reference 20). Characterization of genetic interactions of the isolated mutants established the epistatic relationships between the genes. This is particularly true for *wee1*,[18,19,21] cdc25[22] and cdc2[18,19] that were isolated using the mutants of these screens. From early genetic experiments it became apparent that *wee*1 and *cdc*25 represent opposing forces regulating cdc2 as the *wee*1-50 phenotype is suppressed when combined to cdc25-22.[19] Also overexpression of wild-type cdc25 in a *wee*1-50 background causes cells to undergo a lethal premature mitosis, a phenomenon called mitotic catastrophe.[22]

This convincing genetic data linking *wee*1, *cdc*25, and *cdc*2 was corroborated by biochemical studies leading to the identification of an inhibitory phosphorylation (Tyr-15) on Cdc2 mediated by the Wee1 kinase[23] and dephosphorylated by Cdc25.[24] During these studies a second phosphorylation site was identified on Cdc2 and mapped to Thr-167.[23] It was shown that phosphorylation of this site was critical for Cdc2 activity and stabilized cyclin binding.[25] However, as the critical regulatory steps in *S. pombe* cell cycle progression appear to be cyclin binding and dephosphorylation of Cdc2 by Cdc25 (reviewed in reference 20), the regulatory role of the activating phosphorylation on Thr-167 has remained a puzzle. Yet alanine substitution of Cdc2 Thr-167 is lethal for *S. pombe* and an overexpressed Cdc2 alanine substitution mutant acts in a dominant negative fashion when overexpressed, strongly suggesting that the phosphorylation is essential for Cdc2 function.[25]

Mcs6-Mcs2-Pmh1 is a CDK-Activating Kinase Related to Metazoan CDK7-Cyclin H-MAT1

Thus the concept that CDKs require a phosphorylation in their T-loop for full activity and indeed that such a phosphorylation on Cdc2 is essential was formalized in *S. pombe*. Despite this, the first CDK-activating kinase (CAK) was identified from marine invertebrates and mammalian species as a trimeric complex consisting of the catalytic CDK7 (or MO15) kinase, cyclin H and MAT1 capable of activating Cdc2 in vitro (reviewed by Poon in Chapter 1 of this book).

In retrospect however it has become evident that prior to any knowledge of CAK kinases, genetic analysis of regulators of mitosis in *S. pombe* had revealed two loci termed *Mcs2* and *Mcs6*[27] subsequently identified as key players in Cdc2 activation.

Mcs2 and *Mcs6* were first identified as mutant alleles (*Mcs2-75* and *Mcs6-13*) capable of suppressing a mitotic catastrophe resulting from a hyperactive Cdc2 in the *cdc2-3w wee1-50* double mutant strain. Thus *Mcs2* and *Mcs6* together with four other mitotic catastrophe suppressor (*mcs*) loci were expected to encode positive regulators or substrates of Cdc2.[27] Although most *mcs* loci have been identified by now, only *Mcs2* and *Mcs6* displayed allele-specific interactions with *cdc2* suggesting direct interactions between the encoded proteins.[27] Subsequently Molz and Beach identified *Mcs2* as an essential C-type cyclin,[28] and later *Mcs6* (originally termed crk1 or mop1) was found to encode a CDK associating with Mcs2.[29,30]

Indeed Mcs2 and Mcs6 are the structural homologues of CDK7 and cyclin H respectively.[29,30] The similarity of the *S. pombe* and metazoan complexes include the presence of a homologue of mammalian MAT1[31-34] in *S. pombe.* Thus we recently identified an open reading frame with extensive similarity to MAT1 in sequences released from the *S. pombe* genome sequencing project. The predicted protein consists of 318 amino acids and is encoded by a split gene with five introns designated pmh1 (for pombe MAT1 homologue; GenBank accession no. AF191500). A multiple sequence alignment with MAT1 from *M. musculus,*[34] *X. laevis,*[32] and Tfb3p[35] (also called Rig2p[36] from *S. cerevisiae* (see Fig. 3.1) reveals a strong conservation in the N-terminal part including the RING finger domain. The C-terminus of MAT1 is important for association and activity of the CDK7-cyclin H-MAT1 complex.[37] Interestingly the C-terminus of Pmh1 is more conserved with the metazoan MAT1 than with *S. cerevisiae* Tfb3p, which could reflect differences in the functions of the *S. cerevisiae* Kin28p-Ccl1p-Tfb3p complex compared to *S. pombe* Mcs6-Mcs2-Pmh1 or metazoan CDK7-cyclin H-MAT1 (see below).

The Mcs6 CAK is Activated by the CAK-Activating Kinase (CAKAK) Csk1

The metazoan CDK7-cyclin H dimer can be activated by either association with MAT1 or by phosphorylation of the T-loop threonine (Thr-170 in human CDK7).[34] The identity of the kinase activating this site in metazoans in vivo remains unresolved, but both CDK2 and CDC2 can phosphorylate the site in vitro.[34] *S. pombe* Mcs6 can also be activated by association with Pmh1 (our unpublished results), but like CDK7 it also contains a T-loop phosphorylation site which in the case of Mcs6 is a serine instead of a threonine (Ser-165). First hints regarding the identity of the "CAK-activating kinase" or CAKAK phosphorylating this site were again provided by genetics. During the multicopy suppressor screen to identify *Mcs2*, Molz and Beach[28] identified a second locus capable of rescuing a weak *Mcs2* allele (*Mcs2-75*) that provided a scorable phenotype when combined with *cdc2-3w cdc25-22* alleles. The new locus was termed *Csk1* (for cyclin suppressing kinase), and it encoded a nonessential atypical kinase. Although *Csk1* is not essential, disruption of the gene results in a mild phenotype characterized by a delay of cell cycle re-entry from stationary phase together with abnormalities in the cell wall.[38] *Csk1* is genetically linked not only to *Mcs2* but also to *Mcs6* as demonstrated by the temperature-sensitive synthetic lethality caused by deletion of *Csk1* in combination with the *Mcs6-13* allele.[38] Based on these links it was not a surprise that Csk1 was found to phosphorylate Mcs6 on the T-loop activation site (Ser-165) both in vitro and in vivo.[38] Thus Csk1 constitutes the in vivo CAK of Mcs6, and has therefore also been referred to as a CAK-activating kinase or CAKAK. Interestingly, as *Csk1* is nonessential, the results demonstrate that phosphorylation of the T-loop of Mcs6 is not required for viability[38] as has been observed for the corresponding site in *S. cerevisiae* Kin28p.[39]

The presence of a CAK-activating kinase (CAKAK) adds one additional level of regulation of CDK activity. Indeed through manipulation of Csk1 activity in *S. pombe* one can reciprocally

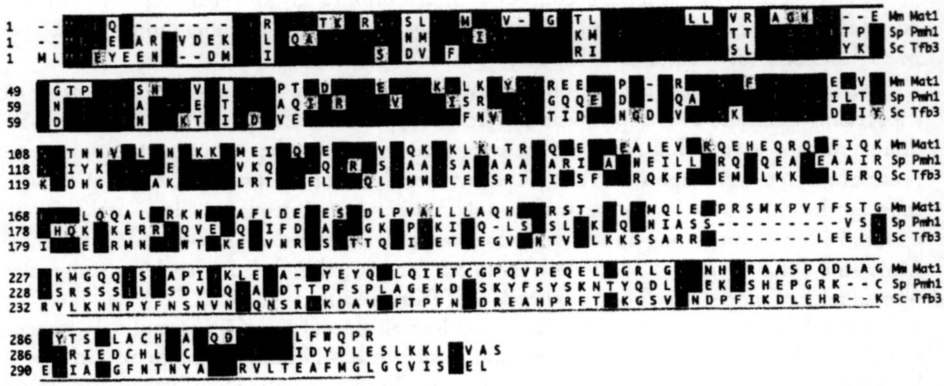

Fig. 3.1. Pmh1 is a MAT1 homologue of S. pombe. Multiple sequence alignment of M. musculus MAT1 (MmMAT1),[34] S. pombe Pmh1 (SpPmh1; Accession no. AF191500) and S. cerevisiae Tfb3p (ScTfb3),[35,36] aligned using Clustal X[53] followed by minor manual alignment of gap-flanking sequences. The N-terminal RING finger (black lines) and the C-terminal CDK7-cyclin H binding domain of metazoan MAT1 (grey lines) are according to Busso et al.[37]

increase or decrease Mcs6 levels and activity in response to either overexpression or deletion of *Csk*1. This suggests the presence of a mechanism to closely monitor and regulate the activity and levels of the Mcs6-Mcs2-Pmh1 complex.[38] This is also reflected in the regulation of Mcs2 levels in a Mcs6 kinase activity-dependent manner (our unpublished results).

Functions of Mcs6-Mcs2-Pmh1 Kinase Complex

Due to the similarities between Mcs6-Mcs2-Pmh1 and CDK7-cyclin H-MAT1[40] as well the ability of Mcs6-Mcs2-Pmh1 to phosphorylate CDKs[29,30] and the C-Terminal Domain (CTD) of the large subunit of RNA polymerase II[29,30] Mcs6-Mcs2-Pmh1 has been implicated in both cell cycle control and in transcription regulation. The proposed transcriptional function has not been addressed in *S. pombe*, although it appears quite likely that the identified CTD kinase activity of *S. pombe* TFIIH[41] will represent Mcs6 as homologues to TFIIH subunits can be readily identified in the *S. pombe* genome.[42,43] If indeed Mcs6-Mcs2-Pmh1 would regulate transcription through CTD phosphorylation, this is likely to represent an essential function of this kinase, and some of the phenotypes described for mutants of *Mcs*2, *Mcs*6, and *Csk*1 may partly reflect this aspect. Of special interest in this regard is the *Csk*1 disruptant phenotype (mimicked by replacement of *Mcs*6 with the alanine substitution mutant (ser-165) of *Mcs*6), which would be consistent with a transcriptional defect.[44] This might suggest that a CTD phosphorylation function of Mcs6 is more sensitive to Mcs6 T-loop phosphorylation than a CAK function.

The role of the Mcs6-Mcs2-Pmh1 complex as a regulator of cell cycle progression on the other hand has been more thoroughly characterized, and there is now compelling evidence indicating that this complex is the CAK of Cdc2 in vivo. The initial identification of both *Mcs*2 and *Mcs*6 as allele-specific interactors of *cdc*2 suggested that these genes were direct upstream (or downstream) regulators of *cdc*2.[28] Here it is interesting to point out that the alleles identified in the thorough screen described above were specifically selected for their ability to suppress a superactive Cdc2. Thus *Mcs*6-13 and *Mcs*2-75 might be selectively defective in their cell cycle function but not in transcription regulation. The single mutation in the *Mcs*6-13 allele (Leu-238 to Arg-238) corresponds to a residue present in subdomain 10 of CDKs and conserved in CDK7 but not in Kin28p.[44] However, the functional consequence of this mutation is difficult to predict, whereas the mutation(s) in *Mcs*2-75 have not been characterized.

In addition to the genetic interactions with *cdc*2, the terminal phenotype in cells from germinating *Mcs*6 null spores appears to arrest as septated cells with condensed chromatin consistent with a mitotic arrest.[29] Indeed this phenotype is very similar for the one reported for *Mcs*2 deletion.[28]

The genetic arguments presented above strongly suggest that Mcs6 is the direct activator of Cdc2 in *S. pombe* and indeed it can activate both Cdc2-Cdc13 and Cdc2-Cig2 complexes in vitro.[45] However, Csk1 was found to activate the same complexes. These studies suggested that either kinase could phosphorylate Cdc2 Thr-167 in vitro, although this was not directly addressed. Furthermore, both overexpressed Mcs6-Mcs2, and Csk1 rescued a *S. cerevisiae* strain with a temperature-sensitive mutation of the single CAK kinase Cak1p of *S. cerevisiae.*[44]

The results from heterologous or in vitro systems thus suggested that Csk1 might activate Cdc2 in addition to Mcs6 also in *S. pombe* cells. This did not appear likely since the *Csk*1 disruption is viable and with normal Cdc2 activity,[28,38,45] but the possibility of redundancy remained. Although the previously described strain combining mutations of both *Csk*1 and *Mcs*6 (*Csk*1Δ *Mcs*6-13) did demonstrate reduced Cdc2 activity[45] this could have been mediated by either or both kinases. To address this issue we recently generated a *S. pombe* strain in which we attempted to combine the effects of both mutations into a single *Mcs*6 allele. We combined the single mutation identified in *Mcs*6-13 (Leu-238 to Arg-238) with the mutation of the Mcs6 T-loop activation site (Ser-165 to Ala-165) blocking Csk1-mediated activation of Mcs6. *S. pombe* cells with this double mutant allele (*Mcs*6-SALR) displayed a temperature-sensitive arrest that was suppressed by *cdc*2.[44] Importantly, the *Mcs*6-SALR strain was not rescued by overexpression of Csk1, although e.g., *S. cerevisiae* Cak1p and human CDK7 did rescue the phenotype. These results strongly suggest that Mcs6 is required and sufficient for CAK activation of Cdc2 in vivo.

Summary and Conclusions

In summary, *S. pombe* has two CAK kinases: Csk1 activating Mcs6 (and thus also a CAKAK) and Mcs6 activating Cdc2. Csk1 is a single subunit kinase distantly related to *S. cerevisiae* Cak1p - indeed these two atypical kinases form a branch of their own in a phylogenetic comparison of CAK kinases.[46] Yet the action of these kinases is distinct as Cak1p also activates the Cdc2 homologue Cdc28p in *S. cerevisiae*. In this respect *S. pombe* cell cycle CDK activation by CAKs appears to be more closely related to metazoans, as in both flies[47] and mice (our unpublished results) the CDK7-associated CAK activity appears to be required for cell cycle progression.

The identification of two CAKs with distinct substrate specificities in a single species is also interesting regarding the activation of the various CDK-cyclin complexes in *S. pombe*. Thus the *S. pombe* genome contains four potential C-type cyclin genes, two of which are uncharacterized but are homologous to *S. cerevisiae* Ctk2p and Srb11p (our unpublished results). The Pch1 C-type cyclin is essential and has been cloned as a yeast two-hybrid partner of Cdc2[48] but the function of this cyclin and its associated kinase activity remains unsolved. In addition, several potential CDK-encoding genes can be identified in the *S. pombe* genome. Beside the Cdc2 and Mcs6 CDKs, Prk1 (homologue of *S. cerevisiae* Srb10p) has been reported[49] and putative sequence homologues belonging to Pho85p, CDK9, Ctk1p and PITSLRE clades have been identified,[46] some of these contain a potential T-loop activation site (our unpublished results). How the various *S. pombe* cyclin-CDK complexes are activated by CAKs remains an open question.

The specificity of CDK activation by CAKs in *S. pombe* is also interesting regarding studies on CAKs in metazoans. In addition to the prototype CDK7-type CAK, CAK activity has been described associated with cyclin K[50] or following affinity-purification of HeLa extracts

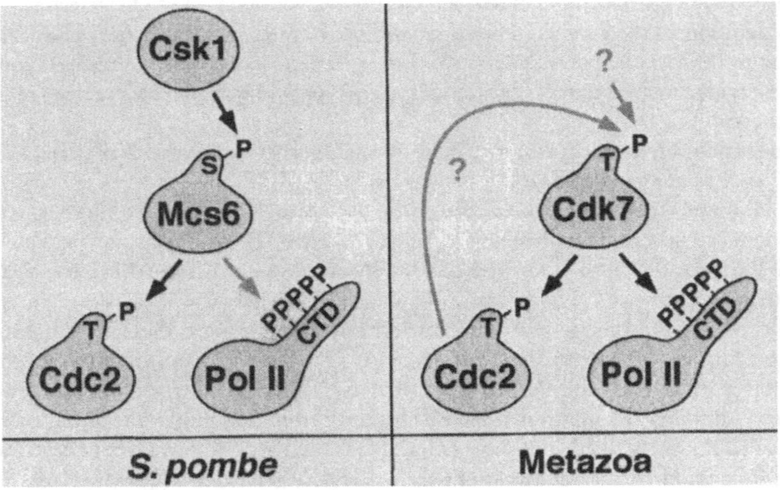

Fig. 3.2. Comparison of CDK-activating kinases in *S. pombe* and metazoa. Schematic presentation of the CDK-activation cascade in *S. pombe*, where the single subunit Csk1 activates Mcs6 which in turn activates Cdc2. The alternative activation of Mcs6 through association with Pmh1 is not shown. Mcs6 is also likely to be involved in phosphorylation of the C-terminal domain (CTD) of the large subunit of RNA polymerase II. The CDK7 kinase complex has been demonstrated to be involved in Cdc2 activation and transcription regulation in flies[47,54] and necessary for cell cycle progression and CTD phosphorylation in mice (our unpublished results). The kinase responsible for CDK7 activation in vivo remains unresolved although both CDK2 and CDC2 can phosphorylate the site in vitro.[34]

with GST-CDK2.[51] The latter strategy revealed a kinase of ca. 30-40 kDa that cross-reacted with Cak1p antisera, and is likely to be identical to a CAK kinase characterized as being downregulated following TGFß treatment.[52] If indeed multiple in vivo CAK kinases are identified also in metazoan species, it will be interesting to study whether one of these might represent a CDK7-activating kinase similar to Csk1 (see Fig. 3.2).

Acknowledgments

We apologize for the studies not reviewed here due to space constraints. We would also like to acknowledge our colleagues and lab members for discussing unpublished work and the entire pombe community for their help and stimulating influence on our work.

References

1. Stern B, Nurse P. A quantitative model for the cdc2 control of S phase and mitosis in fission yeast. Trends Genet 1996; 12(9):345-350.
2. Martin-Castellanos C, Blanco MA, de Prada JM et al. The puc1 cyclin regulates the G1 phase of the fission yeast cell cycle in response to cell size. Mol Biol Cell 2000; 11(2):543-554.
3. Roberts JM. Evolving ideas about cyclins. Cell 1999; 98(2):129-132.
4. Leupold U. Die Vererbung von Homothallie und Heterothallie bei Schizosaccharomyces pombe. C. R. Trav. Lab. Carlsberg, Ser. Physiol. 1950; 24:381-480.
5. Zhao Y, Lieberman HB. Schizosaccharomyces pombe: a model for molecular studies of eukaryotic genes. DNA Cell Biol 1995; 14(5):359-371.

6. Munz P, Wolf K, Kohli J et al. Genetics overview. In: Nasim A, Young P, Johnson BE, eds. Molecular Biology of the Fission Yeast. San Diego: Academic Press, 1989:1-30.
7. MacNeill SA, Fantes PA. Methods for analysis of the fission yeast cell cycle. In: Brooks R, Fantes PA, eds. The Cell Cycle—A Practical Approach. Oxford: IRL Press, 1993:93-125.
8. Hartwell LH, Culotti J, Pringle JR et al. Genetic control of the cell division cycle in yeast. Science 1974; 183(120):46-51.
9. Nurse P, Thuriaux P, Nasmyth K. Genetic control of the cell division cycle in the fission yeast Schizosaccharomyces pombe. Mol Gen Genet 1976; 146(2):167-178.
10. Nurse P, and Thuriaux, P. Control over timing of DNA replication during the cell cycle of fission yeast. Exp. Cell Res. 1977; 107:365-375.
11. Nasmyth K, Nurse P, Fraser RS. The effect of cell mass on the cell cycle timing and duration of S-phase in fission yeast. J Cell Sci 1979; 39:215-233.
12. Nurse P. Cell cycle control genes in yeast. Trends Genet 1985; 1(51):49-55.
13. Connolly T, Beach D. Interaction between the Cigl and Cig2 B-type cyclins in the fission yeast cell cycle. Mol Cell Biol 1994; 14(1):768-776.
14. Mondesert O, McGowan CH, Russell P. Cig2, a B-type cyclin, promotes the onset of S in *Schizosaccharomyces pombe*. Mol Cell Biol 1996; 16(4):1527-1533.
15. Benito J, Martin-Castellanos C, Moreno S. Regulation of the G1 phase of the cell cycle by periodic stabilization and degradation of the $p25^{rum1}$ CDK inhibitor. EMBO J 1998; 17(2):482-497.
16. Mitchison JM. The growth of single cells. I. *Schizosaccharomyces pombe*. Exp Cell Res 1957; 13:244-262.
17. Nurse P. Genetic control of cell size at cell division in yeast. Nature 1975; 256(5518):547-551.
18. Nurse P, Thuriaux P. Regulatory genes controlling mitosis in the fission yeast *Schizosaccharomyces pombe*. Genetics 1980; 96(3):627-637.
19. Fantes PA. Isolation of cell size mutants of a fission yeast by a new selective method: Characterization of mutants and implications for division control mechanisms. J Bacteriol 1981; 146(2):746-754.
20. Fantes PA. Cell cycle controls. In: Nasim A, Young P, Johnson BF, eds. Molecular biology of the fission yeast San Diego: Academic Press 1989: 127-204.
21. Russell P, Nurse P. Negative regulation of mitosis by $wee1^+$, a gene encoding a protein kinase homolog. Cell 1987; 49(4):559-567.
22. Russell P, Nurse P. $cdc25^+$ functions as an inducer in the mitotic control of fission yeast. Cell 1986; 45(1):145-153.
23. Gould KL, Nurse P. Tyrosine phosphorylation of the fission yeast $cdc2^+$ protein kinase regulates entry into mitosis. Nature 1989; 342(6245):39-45.
24. Gould KL, Moreno S, Tonks NK et al. Complementation of the mitotic activator, $p80^{cdc25}$, by a human protein-tyrosine phosphatase. Science 1990; 250(4987):1573-1576.
25. Gould KL, Moreno S, Owen DJ et al. Phosphorylation at Thr167 is required for Schizosaccharomyces pombe $p34^{cdc2}$ function. EMBO J 1991; 10(11):3297-3309.
26. MacNeill SA, Fantes PA. Controlling entry into mitosis in fission yeast. In: Hutchison C, Glover DM, eds. Cell Cycle Control New York: Oxford University Press, 1995:63-105.
27. Molz L, Booher R, Young P et al. cdc2 and the regulation of mitosis: Six interacting mcs genes. Genetics 1989; 122(4):773-782.
28. Molz L, Beach D. Characterization of the fission yeast Mcs2 cyclin and its associated protein kinase activity. EMBO J 1993; 12(4):1723-1732.
29. Buck V, Russell P, Millar JB. Identification of a CDK-activating kinase in fission yeast. EMBO J 1995; 14(24):6173-6183.
30. Damagnez V, Mäkelä TP, Cottarel G. *Schizosaccharomyces pombe* Mop1-Mcs2 is related to mammalian CAK. EMBO J 1995; 14(24):6164-6172.
31. Tassan JP, Jaquenoud M, Fry AM et al. In vitro assembly of a functional human CDK7-cyclin H complex requires MAT1, a novel 36 kDa RING finger protein. EMBO J 1995; 14(22):5608-5617.

32. Devault A, Martinez AM, Fesquet D et al. MAT1 ('menage a trois') a new RING finger protein subunit stabilizing cyclin H-CDK7 complexes in starfish and *Xenopus* CAK. EMBO J 1995; 14(20):5027-5036.
33. Yee A, Nichols MA, Wu L et al. Molecular cloning of CDK7-associated human MAT1, a cyclin-dependent kinase-activating kinase (CAK) assembly factor. Cancer Res 1995; 55(24):6058-6062.
34. Fisher RP, Jin P, Chamberlin HM et al. Alternative mechanisms of CAK assembly require an assembly factor or an activating kinase. Cell 1995; 83(1):47-57.
35. Feaver WJ, Henry NL, Wang Z et al. Genes for Tfb2, Tfb3, and Tfb4 subunits of yeast transcription/repair factor IIH. Homology to human cyclin-dependent kinase activating kinase and IIH subunits. J Biol Chem 1997; 272(31):19319-19327.
36. Faye G, Simon M, Valay JG et al. Rig2, a RING finger protein that interacts with the Kin28/Ccl1 CTD kinase in yeast. Mol Gen Genet 1997; 255(5):460-466.
37. Busso D, Keriel A, Sandrock B et al. Distinct regions of MAT1 regulate CDK7 kinase and TFIIH transcription activities. J Biol Chem 2000; 275(30):22815-22823.
38. Hermand D, Pihlak A, Westerling T et al. Fission yeast Csk1 is a CAK-activating kinase (CAKAK). EMBO J 1998; 17(24):7230-7238.
39. Kimmelman J, Kaldis P, Hengartner CJ et al. Activating Phosphorylation of the Kin28p Subunit of Yeast TFIIH by Cak1p. Mol Cell Biol 1999; 19(7):4774-4787.
40. Kaldis P. The CDK-activating kinase (CAK): from yeast to mammals. Cell Mol Life Sci 1999; 55(2):284-296.
41. Spahr H, Beve J, Larsson T et al. Purification and characterization of RNA polymerase II holoenzyme from Schizosaccharomyces pombe. J Biol Chem 2000; 275(2):1351-1356.
42. Adachi N, Matsumoto M, Hasegawa S et al. Analysis of TFIIH subunit through isolation of the gene from Schizosaccharomyces pombe corresponding to that of Saccharomyces cerevisiae SSL1, reveals the presence of conserved structural motifs. Yeast 1999; 15(3):255-262.
43. Reynolds PR, Biggar S, Prakash L et al. The *Schizosaccharomyces pombe* rhp3$^+$ gene required for DNA repair and cell viability is functionally interchangeable with the RAD3 gene of Saccharomyces cerevisiae. Nucleic Acids Res 1992; 20(9):2327-2334.
44. Hermand D, Westerling T, Pihlak A et al. Specificity of CDK activation in vivo by the two CAKs Mcs6 and Csk1 in fission yeast. EMBO J 2001; 20(1):82-90.
45. Lee KM, Saiz JE, Barton WA et al. Cdc2 activation in fission yeast depends on Mcs6 and Csk1, two partially redundant CDK-activating kinases (CAKs). Curr Biol 1999; 9(8):441-444.
46. Liu J, Kipreos ET. Evolution of cyclin-dependent kinases (CDKs) and CDK-activating kinases (CAKs): Differential conservation of CAKs in yeast and metazoa. Mol Biol Evol 2000; 17(7):1061-1074.
47. Larochelle S, Pandur J, Fisher RP et al. CDK7 is essential for mitosis and for in vivo CDK-activating kinase activity. Genes Dev 1998; 12(3):370-381.
48. Furnari BA, Russell P, Leatherwood J. Pch1$^+$, a second essential C-type cyclin gene in Schizosaccharomyces pombe. J Biol Chem 1997; 272(18):12100-2106.
49. Watson P, Davey J. Characterization of the Prk1 protein kinase from *Schizosaccharomyces pombe.* Yeast 1998; 14(5):485-492.
50. Edwards MC, Wong C, Elledge SJ. Human cyclin K, a novel RNA polymerase II-associated cyclin possessing both carboxy-terminal domain kinase and CDK-activating kinase activity. Mol Cell Biol 1998; 18(7):4291-4300.
51. Kaldis P, Solomon MJ. Analysis of CAK activities from human cells. Eur J Biochem 2000; 267(13):4213-4221.
52. Nagahara H, Ezhevsky SA, Vocero-Akbani AM et al. Transforming growth factor beta targeted inactivation of cyclin E: Cyclin-dependent kinase 2 (CDK2) complexes by inhibition of CDK2 activating kinase activity. Proc Natl Acad Sci U S A 1999; 96(26):14961-14966.
53. Thompson JD, Gibson TJ, Plewniak F et al. The CLUSTAL_X windows interface: Flexible strategies for multiple sequence alignment aided by quality analysis tools. Nucleic Acids Res 1997; 25(24):4876-4882.
54. Leclerc V, Raisin S, Leopold P. Dominant-negative mutants reveal a role for the CDK7 kinase at the mid-blastula transition in *Drosophila* embryos. EMBO J 2000; 19(7):1567-1575.

Chapter 4

Drosophila CDK7:

A Paradigm for CAK in Metazoans

Stéphane Larochelle and Robert P. Fisher

Introduction

Over the last decade, *Drosophila melanogaster* has emerged as a powerful model for the study of cell cycle regulation during development. As is the case in vertebrates, and in contrast to the situation in yeast, the *Drosophila* cell cycle is controlled by multiple cyclin-dependent kinase (CDK) catalytic subunits. To be activated, each catalytic subunit must bind a cyclin partner and undergo phosphorylation on the T-loop by a CDK-activating kinase (CAK). Two types of CAK activity have been identified so far: the monomeric CAKs of budding and fission yeast (Cak1p and Csk1, respectively), and the trimeric, cyclin-containing complexes, exemplified by CDK7/cyclin H/MAT1, found in both vertebrate and invertebrate, metazoan species as well as in fission yeast. *Drosophila* is thus far the only multicellular organism in which a genetic analysis of CDK7 function has been conducted. CDK7 is required in vivo for the activation of CDKs and progression through the cell cycle, and for transcription, as part of the general transcription factor IIH (TFIIH). The results of genetic and biochemical studies of CAK activity in *Drosophila* will be discussed and compared with data from other model systems. From such a comparison, it becomes evident that: 1) the biochemical pathways leading to CDK activation through T-loop phosphorylation have diverged significantly in their details between unicellular and multicellular eukaryotes; and 2) there is a conserved connection between CDK activation and RNA pol II phosphorylation, which is perhaps common to all eukaryotes. The *Drosophila* system, by virtue of its powerful genetics and fundamental conservation with higher eukaryotes, offers perhaps the best opportunity to decipher what this connection means for the coordination of gene expression with cell division in multicellular organisms.

The CDKs of *Drosophila*

Most of the major cell cycle regulatory molecules present in mammalian cells are also found in *Drosophila*. In contrast to yeast, in which a single CDK regulates cell cycle progression through its association with different cyclins, progression through the *Drosophila* cell cycle requires the activation of multiple CDKs. The major CDKs that promote cell cycle progression in *Drosophila* are: CDK1 (also known as CDC2), which forms complexes with cyclins A, B and B3, and is active during mitosis; and CDK2, which binds cyclin E exclusively and functions during the G1/S transition and throughout S phase. *Drosophila* CDK4 is not absolutely required for progression through the cell cycle, but rather seems to be involved in coordinating cell growth and division.[1,2] Although the conservation of CDKs and cyclins between *Droso-*

The CDK-Activating Kinase (CAK), edited by Philipp Kaldis.

phila and vertebrates appears to be the rule, *Drosophila* CDK2 interacts in vivo only with cyclin E,[3] whereas vertebrate CDK2 interacts with both cyclins A and E. The recently completed sequencing of the *Drosophila* genome has also yielded a number of CDK-like molecules for which no functional information is yet available (Table 4.1). Of the 13 CDK or CDK-like genes, mutant phenotypes have been described only for *cdc2*, *cdk2*, *cdk7* and *cdk4*.[1,2,4-7] In fact, most of the requirements inferred for CDK1 and CDK2 during development have been provided by the analyses of their associated cyclins.[3,8-14]

The Cell Cycles of *Drosophila*

One reason why *Drosophila* is an attractive model for the study of cell cycle regulation is that different types of cell cycles can be analyzed in the intact organism. During development, there is a transition from the rapid embryonic cell cycle consisting only of successive S and M phases, to the somatic cycles, which contain gap phases (G1 and G2). Some tissues, moreover, undergo endoreplicative cycles in which S phases follow one another in the absence of mitosis to give rise to polyploid nuclei.[15-17] An exhaustive description of the different types of cell cycles that operate during *Drosophila* development is beyond the scope of this article. However, we will outline some aspects that are helpful in the interpretation of *cdk7* mutant phenotypes and the defects that should result from the loss of T-loop phosphorylation on specific CDKs.

Embryogenesis

Upon fertilization, the embryo begins a series of cycles that are entirely dependent on maternal mRNAs and proteins previously stockpiled in the egg. The large amount of accumulated cell cycle regulators and the absence of growth allow for very rapid embryonic cell cycles consisting of consecutive, synchronous S and M phases that are not separated by any discernible gap phases. The absence of individual cells in the early syncytial embryo obviates the need for cytokinesis and allows for a near-perfect synchrony of nuclear division cycles 1-13. Interphase gradually becomes longer after the tenth mitosis until a G2 phase is finally inserted in cycle 14.[18,19] The transition from maternal to zygotic control of cell division (MZT) occurs after the 13th mitosis, when embryonic transcription is required for the first time.[20,21]

A remarkably thorough analysis of known cell cycle regulators during embryogenesis revealed that different molecular mechanisms govern the cell cycle program during different phases of embryonic development.[22] Although the typical somatic cell cycle is always accompanied by a fluctuation in CDK1 activity—driven by cycles of cyclin synthesis and degradation in concert with tyrosine phosphorylation and dephosphorylation—no such fluctuation can be detected during the first seven embryonic cell cycles.[22] Instead, only a small part of the available pool of cyclin B protein is degraded at each mitosis, while T-loop (Thr-161) phosphorylation of CDK1 appears to be constitutive and Tyr-15 is kept dephosphorylated by String during these cycles.[22] A larger fraction of total cyclin B is degraded in each successive cycle as only cyclin B associated with (the exponentially increasing number of) mitotic spindles is degraded at each anaphase.[23] As a result, CDK1/cyclin B activity is transiently and locally reduced to allow nuclei to return to interphase despite high levels of CDK1 activity in the rest of the cytoplasm.[22,23]

During cycles 8-13, there is still no detectable inhibitory (Tyr-15) phosphorylation on CDK1.[22] Beginning in interphase 9, however, there is an apparent reduction in the level of activating (T-loop) phosphorylation. The fluctuation gradually increases so that by the interphase of cycle 13, the Thr-161-phosphorylated form is undetectable.[22] In addition, oscillation of the total cyclin level can be detected starting in cycle 10. Partly as a result of the depletion of maternally provided cyclin mRNAs, the time required to complete cycles 10 to 13 is lengthened. The activity of CDK1 during cycles 8-13 correlates with the fluctuations in cyclin degradation and T161 phosphorylation. This suggests that cyclin synthesis and degradation

Table 4.1. Drosophila CDKs and their cyclins

CDK	Cyclin	Function	T-loop Phosphorylation?
CDC2 (CDK1)*	A, B, B3	G_2/M, M	Yes
CDK2 (CDC2c)*	EI, EII[a]	G_1/S, S	Yes
CDK4*	D	G_1	Yes
CDK5	p35[b]	Neural development	Putative
CDK7*	H	Transcription (TFIIH) Cell cycle (CAK) all phases	Yes
CDK8	C	Transcription	No
CDK9	T	Transcription (pTEFb)	Putative
Cdkrk	?	?	Putative
PFTAIRE/Eip63E	?	Transcription	Putative
PITSLRE	?	?	Putative
CG6800	?	?	Putative
AC017581	?	?	Putative
AC018104	?	?	Putative

*CDK genes for which mutant phenotypes have been described
[a] Cyclin EI and EII are splice variants of the same gene
[b] p35 is not related to cyclin proteins as define by sequence homology

are major contributors to the regulation of these cycles.[22] It is also possible that the fluctuation in T-loop phosphorylation plays a direct, rate-limiting role in regulating these cycles, although this hypothesis has never been tested. It has likewise not been determined whether the exit from mitosis depends on T-loop dephosphorylation, or solely upon cyclin degradation. Whereas some studies in other systems suggest that the dephosphorylation of the T-loop of CDK1 may be required at some point during the cell cycle,[24,25] more recent data support the opposite conclusion in budding yeast.[26] Interestingly, in a *Drosophila* synchronized embryo, both the activity and level of T-loop phosphorylation of the CDK2/cyclin E complexes remain high during G2 and M phase. Thus downregulation of CDK2/cyclin E is not required to proceed through mitosis during the early embryonic cell cycles.[13]

During the post-blastoderm cycles that follow the MZT, dephosphorylation of the inhibitory Tyr-15 of CDK1 appears to limit entry into mitosis. The embryo cellularizes during cycle 14 and from that point, mitosis is no longer synchronous throughout the organism. Discrete mitotic domains or "pockets" of dividing cells now cycle independently.[19] After the MZT, *string* (the homolog of Cdc25, the CDK1-activating phosphatase) mRNA is transcribed in pulses, and the String protein level fluctuates, peaking at mitosis and disappearing during interphase. Although the developmentally regulated expression of *string* is central to the establishment of patterned cell proliferation during cycles 14-16,[27] other factors such as cyclins and CDK inhibitors also play important regulatory roles in the control of late embryonic cell division.[3,28-30] The role of Tyr-15 dephosphorylation in regulating these cycles is mainly inferred from the pattern of *string* expression.[22,27] The phosphorylation state of bulk CDK1 at this

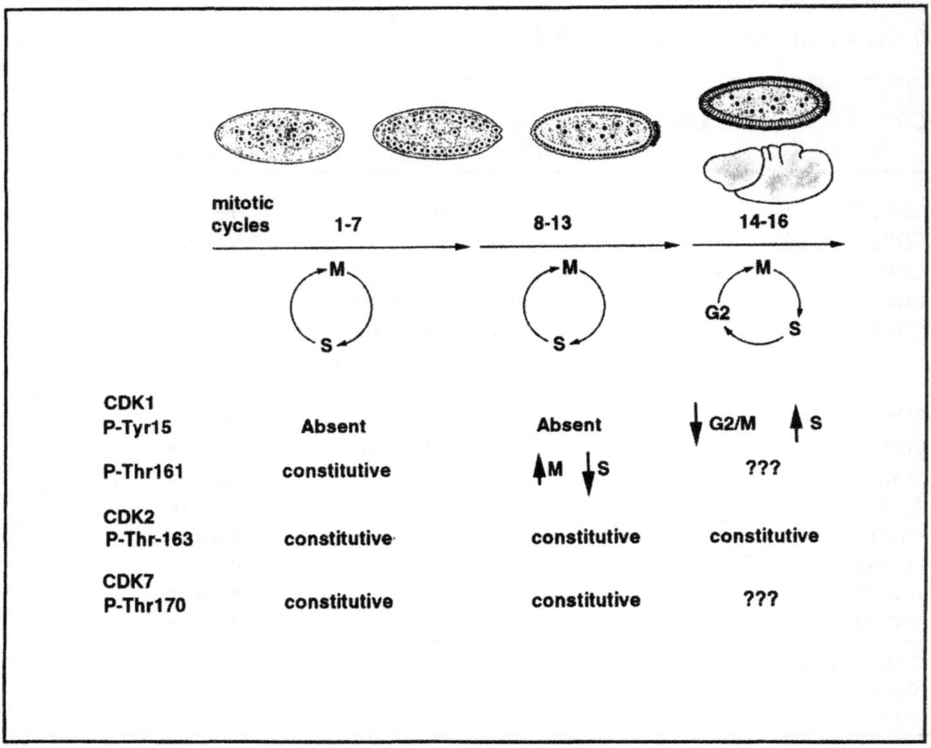

Fig. 4.1. CDK phosphorylation during the embryonic cell cycles. Inhibitory Tyr-15 on CDK1 is absent until the MZT, whereas the activating T-loop phosphorylation on Thr-161 fluctuates starting at cycle 8, with peaks at mitosis.[22] The T-loop of CDK2 remains phosphorylated throughout the cell cycle through cycle 16.[13] Phosphorylation of CDK7 on its T-loop at Thr-170 appears constitutive at least until cycle 14.[93] Diagram adapted from (22, 89).

stage is uninformative, because only a few cells are in mitosis at any one time, and so the potential role of T-loop phosphorylation of CDK1 in regulating these cycles remains elusive.

Mitotic Proliferation and Endoreplication During Larval Development and Oogenesis

After embryogenesis, the animal goes through three larval stages. The bulk of growth during larval development is due to endoreplication, by which polytene cells become larger without increasing in number. Small pockets of mitotically active tissues are present, including imaginal discs, neural cells and a few other tissue types that will later give rise to most of the adult structures.

Because most of the mRNAs required for mitotic proliferation are deposited in large amounts in the egg, their requirement is often masked until late in larval development. The terminal phenotype caused by mutations in many genes specifically required for mitotic proliferation is an outwardly normal looking larva (because larval growth occurs mainly by expansion of tissues undergoing endoreplication) with mitotic tissues that are reduced in size or absent altogether.[31] A good example is *cdk1*. Strong, loss-of-function *cdk1* mutants die at the larval/pupal transition, with rudimentary imaginal discs and brains.[4] Although the polytene cells of the salivary gland appear normal, the imaginal cells found at the neck of this organ are less numerous than in wild-type larva.[4] The analysis of *cdk1* mutants clearly demonstrated a requirement for CDK1 function in mitosis, but not in S phase.[4,32] Further studies with a temperature sensitive allele of *cdk1* also revealed a requirement for CDK1 activity in preventing DNA rereplication before the completion of mitosis in dividing tissues of the larva.[33] Similarly, a requirement for the activity of *cyclin A*, and by extension, CDK1, to prevent ectopic endoreplication had been observed in late embryonic cell cycles.[13]

The *Drosophila* ovary offers several advantages for the study of metazoan cell cycle regulation. First, it is dispensable for viability, and general disruption of cell proliferation can be observed in an otherwise healthy organism, because only the ovarian tissue appears to require continuous mitotic proliferation in the adult. Second, the ovary contains easily distinguishable types of cells undergoing meiosis, mitosis or a modified cell cycle (so-called endo-cycles) resulting in polyploidization through repeated S-phases without intervening mitotic division.[17] Thus, in the adult ovary, several types of cell cycle are easily accessible for study. The phenotypic analysis of mutant ovaries was particularly useful in demonstrating a requirement for CDK7 activity during mitotic proliferation.[6]

A Brief History of CAK

Since the initial observation that CDK1 requires T-loop phosphorylation for its biological function,[34] there has been intense interest in identifying the kinase(s) responsible for this activating phosphorylation.[35,36] Efforts to purify CAK from starfish and frog oocytes resulted in the enrichment of MO15, a protein kinase related to CDKs,[37-39] which had been previously identified in *Xenopus*.[40] MO15 was later renamed CDK7 upon the identification of its activating partner, cyclin H.[41,42] A third subunit, MAT1 ("ménage à trois"), was subsequently identified as a core CAK component.[43-45] In the wake of those discoveries, the budding yeast protein Kin28p, which is ~45% identical to CDK7, seemed a likely candidate to be the budding yeast CAK. Actual experiments, however, soon proved otherwise. Not only did Kin28p fail to work as a CAK in vitro, but its inactivation by mutation in vivo did not cause loss of CDK1 (Cdc28p) T-loop phosphorylation.[46] The loss of viability in *kin28* mutants was instead caused by a virtual shutdown of transcription by RNA pol II.[46] This prompted a new search for CAK in budding yeast by classical purification methods similar to those that had yielded the CDK7 complex in metazoans.[37-39,41,44] These investigations led to the identification of Cak1p (also known as Civ1p), a rather unusual, single-subunit kinase only very distantly related to CDKs.[47-49] Cak1p possessed CAK activity in vitro, and its inactivation in vivo resulted in the loss of CDK1 T-loop phosphorylation, thereby satisfying the criteria for a physiologic CAK in budding yeast.

Those criteria have now also been met in the fission yeast, *Schizosaccharomyces pombe*, albeit with a twist that has yet to be fully explained. Fission yeast contains two CAKs: a trimeric complex containing Mcs6, the ortholog of CDK7;[50-54] and a monomeric kinase, Csk1, which is related to Cak1p.[53-56] Both enzymes can activate CDK1 (Cdc2) in vitro,[54] and it required mutations in both *Mcs6* and *Csk1* to block CDK1 activation and cause a cell cycle arrest in vivo.[54] Because Csk1 activates Mcs6 as well as CDK1 in vitro,[53-55] it may contribute to CDK1 activation only indirectly, as an upstream activator of Mcs6, as has recently been suggested.[57] Further experiments to dissect the functions of Mcs6, which, like its orthologs in other species, functions in transcription as well as in cell cycle control, will be required to determine whether it is the only CDK1-activating kinase in fission yeast. It is clear, nonetheless, that Mcs6, like CDK7 and unlike Kin28p, activates CDKs in vivo.

The Requirement for CAK(s) in *Drosophila*

The importance of CDK7 activity in vivo was first established and explained in *Drosophila*.[6] Although its requirement has not been directly tested in vivo by mutation of the rel-

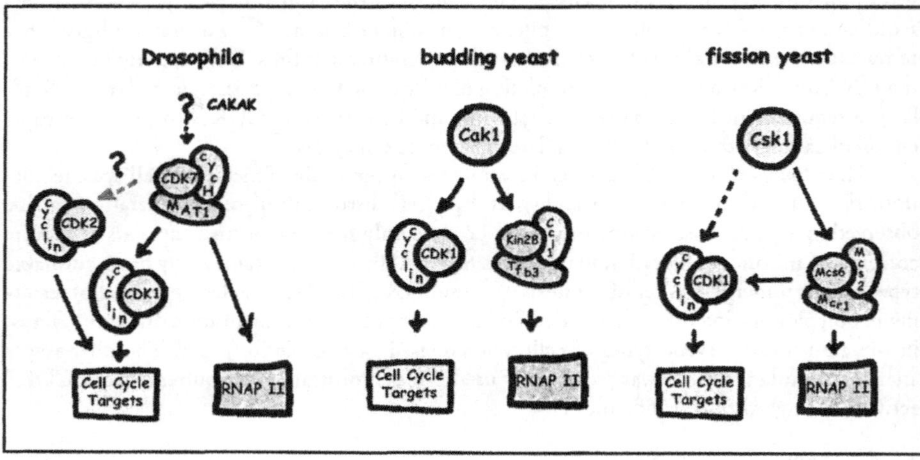

Fig. 4.2. CAK activities in *Drosophila* and yeast. The biochemical pathways leading to CDK activation through T-loop phosphorylation have diverged significantly in their details between unicellular and multicellular eukaryotes. *Drosophila* CDK7 acts both as a CAK[6] and CTD kinase as part of TFIIH,[65] whereas the budding yeast *S. cerevisiae* uses distinct enzymes for those two functions, Kin28p is the kinase subunit of TFIIH, and CAK activity is provided by the monomeric enzyme, Cak1p.[46-49] The fission yeast *S. pombe* has a Cak1p homolog (Csk1) which may also be acting as a CAK for CDK1,[54] although Mcs6 (the CDK7 homolog) has CAK activity in this organism as well.[53,57,90] Cak1p and Csk1 act as upstream activators of Kin28p and Mcs6, respectively.[53-55,91,92] Although the nature of the activity responsible for the T-loop phosphorylation of CDK7 in *Drosophila* (a CAKAK) remains to be determined, such an enzyme could play a key role in regulating the activity of CDK7 towards the CTD of RNA polymerase II.[93] Although it has not yet been demonstrated, it is likely that CDK7 is also responsible for the activation of CDK2 as well as that of CDK1 in *Drosophila*.

evant phosphorylation sites, T-loop phosphorylation on CDK1 and CDK2 can be monitored by the characteristic increase it causes in electrophoretic mobility.[6,13,22] CDK1 and CDK2 in asynchronous *Drosophila* embryos exist predominantly in their unphosphorylated, monomeric forms. In contrast, most cyclin-E bound CDK2 is phosphorylated on its T-loop (at Thr-163) and all of cyclin-A bound CDK1 is phosphorylated within the T-loop (at Thr-161). In these two cases, therefore, either T-loop phosphorylation favors stable interaction with cyclins or, alternatively, cyclin-binding favors T-loop phosphorylation. Cyclin B appears to bind equally well to both phosphorylated and unphosphorylated CDK1.[6] Any impairment of CAK function in *Drosophila* should therefore cause detectable changes in the distribution of CDKs 1 and 2 between different phospho-isoforms, and possibly between cyclin-bound and unbound fractions, as well as cell cycle defects.

Drosophila CDK7, a CAK in Vivo

The *Drosophila cdk7* gene encodes a protein that is ~70% identical to mammalian CDK7. Like its mammalian counterpart, *Drosophila* CDK7 has CAK activity *in vitro*, suggesting that they are functional homologs.[6] The analysis of a *cdk7* null mutation revealed that a complete lack of CDK7 activity is lethal at the larval/pupal transition.[6] The imaginal tissues of $cdk7^{null}$ larvae—the mitotically active progenitors of most adult structures—are substantially smaller or altogether absent, whereas the surrounding polytene tissues appear to be normal.[6] Late larval stage lethality is characteristic of mutations in genes with a large maternal contribution of mRNA, which can provide essential function late into development despite the absence of zygotic transcription. The lack of proliferation of the imaginal tissues, moreover, suggested a role for CDK7 in mitotic proliferation. Several mutations in other *Drosophila* genes thought to be required for mitosis show a similar cell proliferation phenotype.[31] Most significantly, an amorphic allele of the *cdc2* (*cdk1*) gene—the product of which is a target for activation by CAK—has a similar phenotype.[4] The phenotype observed in the $cdk7^{null}$ mutation is therefore consistent with its proposed function as CAK.

Because of its predicted role in transcription, however, impairment of CDK7 function could also block the cell cycle by preventing the transcription of a cell cycle gene, rather than by directly preventing T-loop phosphorylation. The two roles of CDK7 were definitively distinguished by analyzing a conditional (temperature-sensitive) allele of *cdk7* in the mitotically active adult ovary. The most striking phenotype of the $cdk7^{ts}$ mutant transferred to the restrictive temperature was progressive depletion of both the germ cells and the somatically-derived follicle cells of the ovary,[6] strongly suggesting a cell proliferation defect. The identical phenotype was observed with two different, conditional alleles of *cdc2*.[6,32] Given the ability of CDK7 to activate CDK1 in vitro, and their identical phenotypes when inactivated,[6,32] the specific mitotic defect caused by the $cdk7^{P140S}$ allele is likely to be a direct consequence of CAK failure.

It is, on the other hand, unlikely that the phenotypes observed in the $cdk7^{P140S}$ mutant resulted from a general transcription defect, for two reasons. First, DNA replication in the germ cells continued for days after the complete arrest of cell division.[6] Because the repeated S phases of the endoreplicating nuclei depend on the periodic transcriptional induction of cyclin E,[8,14] it seems this particular transcription program, at least, has not been altered in the $cdk7^{P140S}$ mutant. Second, any global reduction in RNA pol II transcription would be expected to shorten life span, whereas the viability of the adult $cdk7^{P140S}$ animals is not compromised at high temperature. In contrast, rapid temperature-sensitive lethality does occur in adult flies with the more severely defective allele, $cdk7^{S164A/T170A}$.[93] This difference argues that the P140S mutation spares the essential transcription function of CDK7, even at high temperatures.

Biochemical analyses of CDK-cyclin complex formation and activity in the $cdk7^{P140S}$ embryos essentially confirmed a deficiency of CDK1-activating kinase in vivo. Cyclin B-bound CDK1 isolated from $cdk7^{ts}$ embryos that were collected at the restrictive temperature is hypophosphorylated on its T-loop and exhibits low kinase activity. Full activity can be restored to the CDK1/cyclin B complex by treatment with recombinant CDK7-cyclin H,[6] indicating that the only impediment to kinase activation is the reduction in T-loop phosphorylation. In contrast to the CDK1/cyclin B and CDK2/cyclin A complexes, the mammalian CDK1/cyclin A complex cannot form in the absence of T-loop phosphorylation.[58] The same appears to hold true in *Drosophila*, because only the T-loop-phosphorylated form of CDK1 can be found in cyclin A immunoprecipitates from wild type embryos.[6] Consistent with this model, inactivation of *cdk7* results in the failure of cyclin A to bind CDK1.[6] Finally, all measurable CAK activity in an embryonic extract can be immunodepleted with a CDK7-specific antibody, indicating that CDK7 is the major, if not the only CAK activity present in *Drosophila* embryos.[6] In a similar manner, all CAK activity can be depleted from mammalian and *Xenopus* oocyte extracts with antibodies against CDK7.[59,60] Therefore, if there is another CAK in these extracts, it is either present in vanishingly small quantities undetectable by conventional methods, or inactive under standard assay conditions in vitro.

The $cdk7^{P140S}$ allele was designed by analogy with known, temperature-sensitive mutant forms of both yeast and *Drosophila* protein kinases.[5,61,62] The $CDK7^{P140S}$ protein, however, is not itself temperature-sensitive in vitro. The likely explanation for the temperature-sensitive phenotype, therefore, is the slow depletion of $CDK7^{P140S}$ protein over a period of days at the restrictive-temperature.[6] The slow kinetics of CDK7 protein disappearance may mean that only protein synthesized at the restrictive temperature is quickly degraded, while the CDK7 protein previously synthesized at the permissive temperature remains functional until it is turned over normally.[6] Such temperaturesensitive folding (*tsf*) mutants, which are active at high temperature when synthesized at low temperature,[63,64] tend to be stabilized by association with other subunits within a complex. Because CDK7 is part of (at least) two distinct multi-subunit complexes, the $CDK7^{P140S}$ protein may be stabilized following its incorporation into either free CAK or TFIIH complexes. The selective loss of CDK1 activation without a general transcription defect in the $cdk7^{P140S}$ mutant might therefore reflect a slower rate of CDK7 turnover in the TFIIH complex relative to the free CAK complex.

If CDK7 is responsible for rephosphorylating the T-loop of CDK1 after embryonic interphase 9, one would predict that its rapid inactivation caused by the expression of a dominant negative form of CDK7 in the early syncytial blastoderm would lead to a cell cycle arrest.[65] The requirement for the CAK function could, in theory, be observed independent of transcription, which is not required during cycles 1-13. When such a manipulation was attempted, however, no effect was observed either on cell cycle progression or on CDK1 T-loop phosphorylation.[65] It is unfortunately difficult to draw conclusions from the negative results obtained by this approach. Neither of the *cdk7* alleles introduced in these studies actually corresponds to the "classical" dominant-negative CDK mutation—an Asp-to-Asn mutation in the "DFG" motif of subdomain VII—that abrogates catalytic activity without affecting cyclin-binding.[66] The maternally contributed CDK7/cyclin H/MAT1 complex is very abundant in the early embryo, and once formed, is probably quite resistant to subunit exchange, even with versions of the kinase that could be predicted to bind efficiently to cyclin H and MAT1. Indeed, the measured reduction in CDK7-associated CAK activity ranged from 50% to 75% upon overexpression of either $CDK7^{D137R}$ (Asp-to-Arg in subdomain VIb[67]) or $CDK7^{T170A}$ (Thr-to-Ala in the T-loop) mutant proteins.[65] Because the expression of the mutant transgenes was localized to the posterior pole, it was argued that inhibition of CDK7 activity in that region might be greater.[65] However, CDK7 activity can be reduced to less than 4% of wild type levels with no detectable phenotypic consequence.[93] Moreover, although developmental defects were reported in animals overexpressing $CDK7^{T170A}$,[65] expression of the same variant at near wild-type levels in an otherwise $cdk7^{null}$ background results in phenotypically wild-type animals at all stages of development, and supports CAK activity in extracts that is comparable to that of wild-type CDK7.[93] At this point, therefore, the role of CDK7 in regulating embryonic cycles 9-13 remains uncertain, because untested.

Drosophila CDK7 in Transcription

It became apparent that CDK7 and its orthologs, Kin28p and Mcs6, have an important role in transcription when CDK7 was identified as the kinase subunit of the general transcription factor IIH (TFIIH).[68-70] A mutation in the budding yeast *KIN28* gene globally impairs transcription by RNA pol II.[46,71,72] In association with TFIIH, CDK7 phosphorylates the carboxyl-terminal domain (CTD) of RNA polymerase II (RNA Pol II) to facilitate promoter clearance.[73] Although some of the previously described developmental defects in *Drosophila* loss-of-function mutations are likely to reflect compromised transcription functions, they were not characterized at the molecular level.[6] More recently, effects of interfering with CDK7 function at various times during *Drosophila* development were analyzed by expressing different mutant forms of CDK7[65] and observing dramatic derangements of wing and eye development perhaps consistent with the suspected role of CDK7 in transcription.[65] As noted above, however, those phenotypes probably reflect to a significant degree the overexpression of mutant CDK7 proteins, rather than simple, quantitative impairment of CDK7 function.

We have recently characterized an allele of *Drosophila cdk7* that causes temperaturesensitive lethality at all stages of development, and death to adult flies shifted to 29°C after 48-72 hours.[93] Because cell division is not required in adult flies, this almost certainly reflects the general requirement for TFIIH-associated CDK7 activity in transcription by RNA pol II. Analysis of a more subtle mutation gives some insight into the mechanism by which transcription, and perhaps viability, is impaired when CDK7 function is compromised. The $CDK7^{T170A}$ mutant protein cannot be phosphorylated at its own activating threonine residue, and as a result is kinetically impaired in its ability to phosphorylate the CTD, with activity ~25-fold lower than that of the wild-type enzyme.[93] Interestingly, CAK activity is not affected by the mutation.[93] Within the context of a preinitiation complex, TFIIH-associated CDK7 phosphorylates the CTD in opposition to CTD phosphatases.[74] Thus T-loop phosphorylation of CDK7 provides a kinetic mechanism whereby the balance between kinase and phosphatase activity can be "switched" to modulate the CTD phosphorylation state in response to environmental and developmental signals. In third instar larvae subjected to a brief heat shock, for example, the CTD of RNA pol II becomes transiently hyperphosphorylated; the increased rate of CTD modification is blunted or abrogated in the $cdk7^{T170A}$ mutant.[93]

Regulation of CDK7 Activity

How is the activity of CDK7 itself regulated, if at all? All attempts to uncover fluctuations in the level of CAK activity during the cell cycle have failed.[75,76] It remains possible that CDK7 activity is regulated in a manner that is not reflected by measurements of total, extractable CDK7. It has been reported that the substrate specificity of CDK7 changes depending on the presence or absence of MAT1,[77] and that MAT1 seems to be required for the efficient phosphorylation of a number of substrates in vitro.[75-80] There is little evidence, however, that large amounts of MAT1-free, dimeric CDK7/cyclin H exist in mammalian cells and it is thus doubtful that the association with MAT1 serves as a regulatory mechanism in vivo. Like other CDKs, CDK7 is phosphorylated within its T-loop on a conserved threonine residue, Thr-170.[44,75,81,82] In addition, a serine residue (Ser-164) within the T-loop is the target of phosphorylation in vivo and in vitro.[44,81,82] Although neither phosphorylation is required for viability in *Drosophila* under normal laboratory conditions, the complete absence of phosphorylation on the T-loop of CDK7 results in lethality at temperatures above 24°C.[93] Moreover, as described above, the phosphorylation of Thr-170 uniquely stimulates the CTD kinase activity of the CDK7/cyclin H/MAT1 complex without appreciably affecting CAK activity, and could therefore be involved in regulating the rate of transcription by RNA pol II.[93] Whether there are regulatory modalities specific to the CAK function of the CDK7 complex remains to be tested.

Evolutionary Considerations

Most of the enzymes involved in regulating cell cycle progression are conserved from yeast to humans. Structural orthologs are often, although not always, functional homologs. When there is divergence of the two, it is often attributable to the expansion of gene families in rough proportion to an organism's complexity. For example, a single CDK drives cell cycle progression in yeast, whereas vertebrates rely on several distinct CDKs. Hence, the surprise when it was shown that two functions requiring two different enzymes in budding yeast are both carried out in vivo by metazoan CDK7. Clearly, however, molecular complexity does not always correlate in obvious ways with organismal complexity: budding yeast CDK1, for example, associates with a larger variety of cyclins than does its mammalian counterpart.[83]

In spite of considerable evidence that CDK7 is a true CAK *in vivo*, skepticism about this point continues to pervade the CAK literature.[35,65,84] One result often cited to support the existence of additional or alternative CAKs in metazoans is the failure to observe loss of CDK2 T-loop phosphorylation when CDK7 is inactivated in *Drosophila*.[6] It has never been demonstrated, however, that the T-loop phosphorylation of CDK2 fluctuates during the cell cycle,[13] perhaps indicating that the T-loop of CDK2 is not normally dephosphorylated between rounds of activation by cyclin E. It also remains possible that, even if CDK2 T-loop phosphorylation normally fluctuates, the arrest caused by the *cdk7*^{P140S} allele occurs prior to the point at which CDK2 dephosphorylation normally occurs.[6] Finally, failure to activate CDK1 was observed with a single allele of *cdk7*, which is unlikely to cause a complete loss of function (see above), and thus might be able to provide CAK activity in vivo sufficient to maintain T-loop phosphorylation of CDK2.

Are there Other CAKs in *Drosophila*?

It was anticipated that the availability of complete genome sequences would finally reveal the elusive *CAK1* homologs of such metazoans as *Drosophila* and *C. elegans*,[56] but this was not the case. Similarly, no obvious *CAK1* homolog could be identified within the human genome.[85] This suggests that, even if another protein kinase capable of acting as a CAK does exist in higher eukaryotes, it shares little sequence homology with yeast CAKs. This would not be a complete surprise; Cak1p and Csk1 share only 20-25% identity, but appear to be orthologs, based on both cross-complementation studies[57,86] and phylogenetic analyses.[56] Similarly the *C. albicans* Cak1p homolog shares the same low homology with its *S. cerevisiae* counterpart.[86] It therefore remains possible that a *Drosophila* kinase similar to Cak1p or Csk1 exists but cannot be clearly identified on the basis of sequence homology alone. What such an enzyme might be doing in vivo is another question. A potentially satisfying, but not necessarily likely possibility is a specialized function in activating CDK2 (see above). Another intriguing possibility, with perhaps a precedent in the fission yeast system,[53-55,57] would be a Cak1p/Csk1-like enzyme specific for CDK7. Finally, in the context of a multicellular organism, there is the possibility that a metazoan Cak1p has evolved to perform a tissue-specific CAK function.

Conclusions

The *Drosophila* system has provided the best evidence to date that CDK7 is a bona fide, physiologic CAK in metazoans. Furthermore, *Drosophila* may serve as a model system to explore other aspects of CAK function and regulation that have yet to be elucidated. For example, other kinases capable of activating CDKs, perhaps similar to one recently identified in HeLa cells,[86] may be identified in *Drosophila*. The division of labor between such an enzyme and CDK7 could be analyzed at different stages of development, with the powerful genetic tools available in *Drosophila*.

Dissecting the requirements for T-loop phosphorylation of CDKs by CAK will be equally interesting. In particular, little is known about whether this modification plays an important role in regulating cell division during development. Thus far, the only *Drosophila* cell cycles in which T-loop phosphorylation of a CDK is known to fluctuate are the early embryonic cell cycles.[22] It remains to be tested thoroughly whether T-loop phosphorylation of CDK1 actually limits the rate of entry into mitoses 9-13. The rapid nuclear cycles of the early embryo cannot be taken as a general representation of the somatic cell cycle. Regulation at one developmental stage would, however, prompt a renewed search for other contexts in which T-loop phosphorylation is an actual, rather than merely a potential, regulator of cell cycle progression.

Finally, given the dual function of CDK7 in promoting cell cycle progression as well as transcription by RNA pol II, *Drosophila* will be a useful model to study what may be the most critical question to emerge from studies of CAKs in all eukaryotes: how those two functions are coordinated during development. Although different organisms have different strategies for phosphorylating the T-loops of CDKs, all species maintain a close connection between this event and phosphorylation of the CTD of the RNA pol II large subunit. The notion that CAK/TFIIH-associated kinase represents a critical intersection of multiple metabolic pathways,[87,88] integrating and coordinating signals important for cell proliferation, gene expression and genomic integrity, remains to be explored systematically. The *Drosophila* system has the right combination of features—powerful genetics, sophisticated cell and developmental biology, and essential conservation with vertebrate systems—to promote such future explorations.

Acknowledgments

We thank Reiko Cyr for critical reading of the manuscript and support. Work in the authors' laboratory is funded by grants to R.P.F. from the National Institute of Health and the American Cancer Society. S.L. is a Research Fellow of the National Cancer Institute of Canada with funds provided by the Terry Fox run.

References

1. Datar SA, Jacobs HW, de La Cruz AF, Lehner CF, Edgar BA. The *Drosophila* cyclin D-cdk4 complex promotes cellular growth. EMBO J 2000; 19:4543-4554.
2. Meyer CA, Jacobs HW, Datar SA, Du W, Edgar BA et al. *Drosophila* cdk4 is required for normal growth and is dispensable for cell cycle progression. EMBO J 2000; 19:4533-4542.
3. Knoblich JA, Sauer K, Jones L, Richardson H, Saint R et al. Cyclin E controls S phase progression and its down-regulation during *Drosophila* embryogenesis is required for the arrest of cell proliferation. Cell 1994; 77:107-120.
4. Stern B, Ried G, Clegg NG, Grigliatti TA, Lehner CF. Genetic analysis of the *Drosophila* cdc2 homolog. Development 1993; 117:219-232.
5. Sigrist S, Ried G, Lehner CF. Dmcdc2 kinase is required for both meiotic divisions during *Drosophila* spermatogenesis and is activated by the Twine/cdc25 phosphatase. Mech Dev 1995; 53:247-250.
6. Larochelle S, Pandur J, Fisher RP, Salz HK, Suter B. Cdk7 is essential for mitosis and for in vivo CDK-activating kinase activity. Genes Dev 1998; 12:370-381.
7. Lane ME, Elend M, Heidmann D, Herr A, Marzodko S et al. A screen for modifiers of cyclin E function in *Drosophila* melanogaster identifies Cdk2 mutations, revealing the insignificance of putative phosphorylation sites in Cdk2. Genetics 2000; 155:233-244.
8. Follette PJ, Duronio RJ, O'Farrell PH. Fluctuations in cyclin E levels are required for multiple rounds of endocycle S phase in *Drosophila*. Curr Biol 1998; 12:235-238.
9. Jacobs HW, Knoblich JA, Lehner CF. *Drosophila* cyclin B3 is required for female fertility and is dispensable for mitosis like cyclin B. Genes Dev 1998; 12(23):3741-3751.
10. Lehner CF, O'Farrell PH. Expression and function of *Drosophila* cyclin A during embryonic cell cycle progression. Cell 1989; 56(6):957-968.
11. Lehner CF, O'Farrell PH. The roles of *Drosophila* cyclins A and B in mitotic control. Cell 1990; 61:535-547.
12. Richardson H, O'Keefe LV, Marty T, Saint R. Ectopic cyclin E expression induces premature entry into S phase and disrupts pattern formation in the *Drosophila* eye imaginal disc. Development 1995; 121(10):3371-3379.
13. Sauer K, Knoblich JA, Richardson H, Lehner CF. Distinct modes of cyclin E/cdc2c kinase regulation and S-phase control in mitotic and endoreduplication cycles of *Drosophila* embryogenesis. Genes Dev 1995; 9:1327-1339.
14. Weiss A, Herzig A, Jacobs H, Lehner CF. Continuous cyclin E expression inhibits progression through endoreduplication cycles in *Drosophila*. Curr Biol 1998; 12:239-242.
15. Edgar BA, Lehner CF. Developmental control of cell cycle regulators: A fly's perspective. Science 1996; 274:1646-1652.
16. Vidwans SJ, Su TT. Cycling through development in *Drosophila* and other metazoa. Nat Cell Bio 2001; 3:E35-E39.
17. Edgar BA, Orr-Weaver TL. Endoreplication cell cycles: More for less. Cell 2001; 105:297-306.
18. Edgar BA, O'Farrell PH. The three postblastoderm cell cycles of *Drosophila* embryogenesis are regulated in G2 by string. Cell 1990; 62(3):469-480.
19. Foe VE. Mitotic domains reveal early commitment of cells in *Drosophila* embryos. Development 1989; 107(1):1-22.
20. Edgar BA, Schubiger G. Parameters controlling transcriptional activation during early *Drosophila* development. Cell 1986; 44(6):871-877.

21. Edgar BA, Datar SA. Zygotic degradation of two maternal Cdc25 mRNAs terminates *Drosophila*'s early cell cycle program. Genes Dev 1996; 10(15):1966-1977.
22. Edgar BA, Sprenger F, Duronio RJ, Leopold P, O'Farrell PH. Distinct molecular mechanisms regulate cell cycle timing at successive stages of *Drosophila* embryogenesis. Genes Dev 1994; 8:440-452.
23. Su TT, Sprenger F, DiGregorio PJ, Campbell SD, O'Farrell PH. Exit from mitosis in *Drosophila* syncytial embryos requires proteolysis and cyclin degradation, and is associated with localized dephosphorylation. Genes Dev 1998; 12(10):1495-1503.
24. Lim HH, Loy CJ, Zaman S, Surana U. Dephosphorylation of threonine 169 of Cdc28 is not required for exit from mitosis but may be necessary for Start in *Saccharomyces cerevisiae*. Mol Cell Biol 1996; 16:4573-4583.
25. Lorca T, Labbé J-C, Devault A, Fesquet D, Capony J-P et al. Dephosphorylation of cdc2 on threonine 161 is required for cdc2 kinase inactivation and normal anaphase. EMBO J 1992; 11:2381-2390.
26. Cross FR, Levine K. Genetic analysis of the relationship between activation loop phosphorylation and cyclin binding in the activation of the Saccharomyces cerevisiae Cdc28p cyclin-dependent kinase. Genetics 2000; 154(4):1549-1559.
27. Edgar BA, Lehman DA, O'Farrell PH. Transcriptional regulation of string (cdc25): A link between developmental programming and the cell cycle. Development 1994; 120(11):3131-3143.
28. Richardson HE, O'Keefe LV, Reed SI, Saint R. A *Drosophila* G1-specific cyclin E homolog exhibits different modes of expression during embryogenesis. Development 1993; 119:673-690.
29. Lane ME, Sauer K, Wallace K, Jan YN, Lehner CF et al. Dacapo, a cyclin-dependent kinase inhibitor, stops cell proliferation during *Drosophila* development. Cell 1996; 87(7):1225-1235.
30. de Nooij JC, Letendre MA, Hariharan IK. A cyclin-dependent kinase inhibitor, Dacapo, is necessary for timely exit from the cell cycle during *Drosophila* embryogenesis. Cell 1996; 87(7):1237-1247.
31. Gatti M, Baker BS. Genes controlling essential cell-cycle functions in *Drosophila melanogaster*. Genes Dev 1989; 3(4):438-453.
32. Reed BH, Orr-Weaver TL. The *Drosophila* gene *morula* inhibits mitotic functions in the endo cell cycle and the mitotic cell cycle. Development 1997; 124:3543-3553.
33. Hayashi S. A Cdc2 dependent checkpoint maintains diploidy in *Drosophila*. Development 1996; 122:1051-1058.
34. Gould KL, Moreno S, Owen DJ, Sazer S, Nurse P. Phosphorylation at Thr167 is required for *Schizosaccharomyces pombe* $p34^{cdc2}$ function. EMBO J 1991; 10:3297-3309.
35. Harper JW, Elledge SJ. The role of Cdk7 in CAK function, a retro-retrospective. Genes Dev 1998; 12:285-289.
36. Kaldis P. The CDK-activating kinase (CAK): From yeast to mammals. Cell Mol Life Sci 1999; 55:284-296.
37. Fesquet D, Labbé J-C, Derancourt J, Capony J-P, Galas S et al. The *MO15* gene encodes the catalytic subunit of a protein kinase that activates cdc2 and other cyclin-dependent kinases (CDKs) through phosphorylation of Thr161 and its homologues. EMBO J 1993; 12:3111-3121.
38. Poon RYC, Yamashita K, Adamczewski JP, Hunt T, Shuttleworth J. The cdc2-related protein $p40^{MO15}$ is the catalytic subunit of a protein kinase that can activate $p33^{cdk2}$ and $p34^{cdc2}$. EMBO J 1993; 12:3123-3132.
39. Solomon MJ, Harper JW, Shuttleworth J. CAK, the $p34^{cdc2}$ activating kinase, contains a protein identical or closely related to $p40^{MO15}$. EMBO J 1993; 12:3133-3142.
40. Shuttleworth J, Godfrey R, Colman A. $p40^{MO15}$, a cdc2-related protein kinase involved in negative regulation of meiotic maturation in Xenopus oocytes. EMBO J 1990; 9:3233-3240.
41. Fisher RP, Morgan DO. A novel cyclin associates with MO15/CDK7 to form the CDK-activating kinase. Cell 1994; 78:713-724.
42. Mäkelä TP, Tassan J-P, Nigg EA, Frutiger S, Hughes GJ et al. A cyclin associated with the CDK-activating kinase MO15. Nature 1994; 371:254-257.

43. Devault A, Martinez A-M, Fesquet D, Labbé J-C, Morin N et al. MAT1 (*'ménage à trois'*) a new RING finger protein subunit stabilizing cyclin H-cdk7 complexes in starfish and *Xenopus* CAK. EMBO J 1995; 14:5027-5036.
44. Fisher RP, Jin P, Chamberlin HM, Morgan DO. Alternative mechanisms of CAK assembly require an assembly factor or an activating kinase. Cell 1995; 83:47-57.
45. Tassan J-P, Jaquenod M, Fry AM, Frutiger S, Hughes G et al. *In vitro* assembly of a functional human cdk7/cyclin H complex requires MAT1, a novel 36 kD RING finger protein. EMBO J 1995; 14:5608-5617.
46. Cismowski MJ, Laff GM, Solomon MJ, Reed SI. *KIN28* encodes a C-terminal domain kinase that controls mRNA transcription in *Saccharomyces cerevisiae* but lacks cyclin-dependent kinase-activating kinase (CAK) activity. Mol Cell Biol 1995; 15(6):2983-2992.
47. Espinoza FH, Farrell A, Erdjument-Bromage H, Tempst P, Morgan DO. A cyclin-dependent kinase-activating kinase (CAK) in budding yeast unrelated to vertebrate CAK. Science 1996; 273:1714-1717.
48. Kaldis P, Sutton A, Solomon MJ. The CDK-activating kinase (CAK) from budding yeast. Cell 1996; 86:553-564.
49. Thuret J-Y, Valay J-G, Faye G, Mann C. Civ1 (CAK in vivo), a novel CDK-activating kinase. Cell 1996; 86:565-576.
50. Molz L, Booher R, Young P, Beach D. *cdc2* and the regulation of mitosis: six interacting *mcs* genes. Genetics 1989; 122:773-782.
51. Damagnez V, Mäkelä TP, Cottarel G. *Schizosaccharomyces pombe* Mop1-Mcs2 is related to mammalian CAK. EMBO J 1995; 14:6164-6172.
52. Buck V, Russell P, Millar JBA. Identification of a CDK-activating kinase in fission yeast. EMBO J 1995; 14:6173-6183.
53. Hermand D, Pihlak A, Westerling T, Damagnez V, Vandenhaute J et al. Fission yeast Csk1 is a CAK-activating kinase (CAKAK). EMBO J 1998; 17:7230-7238.
54. Lee KM, Saiz JE, Barton WA, Fisher RP. Cdc2 activation in fission yeast depends on Mcs6 and Csk1, two partially redundant CDK-activating kinases CAKs). Curr Biol 1999; 9:441-444.
55. Molz L, Beach D. Characterization of the fission yeast *Mcs2* cyclin and its associated protein kinase activity. EMBO J 1993; 12:1723-1732.
56. Liu J, Kiperos ET. Evolution of Cyclin-dependent kinases (CDKs) and CDK-activating kinases (CAKs): Differential Conservation of CAKs in Yeast and Metazoa. Mol Biol Evol 2000; 17:1061-1074.
57. Hermand D, Westerling T, Pihlak A, Thuret J-Y, Vallenius T et al. Specificity of cdk activation *in vivo* by the two CAKs Mcs6 and Csk1 in fission yeast. EMBO J 2001; 20:82-90.
58. Desai D, Wessling HC, Fisher RP, Morgan DO. The effect of phosphorylation by CAK on cyclin binding by CDC2 and CDK2. Mol Cell Biol 1995; 15:345-350.
59. Matsuoka M, Kato J, Fisher RP, Morgan DO, Sherr CJ. Activation of cyclin-dependent kinase-4 (CDK4) by mouse MO15-associated kinase. Mol Cell Biol 1994; 14:7265-7275.
60. Fesquet D, Morin N, Doree M, Devault A. Is cdk7/cyclin H/MAT1 the genuine cdk activating kinase in cycling xenopus egg extracts? Oncogene 1997; 15:1303-1307.
61. Carr AM, MacNeill SA, Hayles J, Nurse P. Molecular cloning and sequence analysis of mutant alleles of the fission yeast *cdc2* protein kinase gene: Implications for *cdc2+* protein structure and function. Mol Gen Genet 1989; 218:41-49.
62. Hsu J-C, Perrimon N. A temperaturesensitive MEK mutation demonstrates the conservation of the signaling pathways activated by receptor tyrosine kinases. Genes Dev 1994; 8:2176-2187.
63. Goldenberg DP, King J. Temperaturesensitive mutants blocked in the folding or subunit of the bacteriophage P22 tail spike protein. II. Active mutant proteins matured at 30 degrees C. J Mol Biol 1981; 145(4):633-651.
64. Smith DH, King J. Temperaturesensitive mutants blocked in the folding or subunit assembly of the bacteriophage P22 tail spike protein. III. Intensive polypeptide chains synthesized at 39 degrees C. J Mol Biol 1981; 145(4):653-676.
65. Leclerc V, Raisin S, Léopold P. Dominant-negative mutants reveal a role for the Cdk7 kinase at the mid-blastula transition in *Drosophila* embryos. EMBO J 2000; 19:1567-1575.

66. van den Heuvel S, Harlow E. Distinct roles for cyclin-dependent kinases in cell cycle control. Science 1993; 262:2050-2054.
67. Hanks SK, Quinn AM. Protein kinase catalytic domain sequence database: Identification of conserved features of primary structure and classification of family members. Methods Enzymol 1991; 200:38-62.
68. Roy R, Adamczewski JP, Seroz T, Vermuelen W, Tassan J-P et al. The MO15 cell cycle kinase is associated with the TFIIH transcription-DNA repair factor. Cell 1994; 79:1093-1101.
69. Serizawa H, Mäkelä TP, Conaway JW, Conaway RC, Weinberg RA et al. Association of CDK-activating kinase subunits with transcription factor TFIIH. Nature 1995; 374:280-282.
70. Shiekhattar R, Mermelstein F, Fisher RP, Drapkin R, Dynlacht B et al. CDK-activating kinase (CAK) complex is a component of human transcription factor IIH. Nature 1995; 374:283-287.
71. Holstege FCP, Jennings EG, Wyrick JJ, Lee TI, Hengartner CJ et al. Dissecting the regulatory circuitry of a eukaryotic genome. Cell 1998; 95:717-728.
72. Hengartner CJ, Myer VE, Liao SM, Wilson CJ, Koh SS et al. Temporal regulation of RNA polymerase II by Srb10 and Kin28 cyclin-dependent kinases. Mol Cell 1998; 2(1):43-53.
73. Dahmus ME. Reversible phosphorylation of the C-terminal domain of RNA polymerase II. J Biol Chem 1996; 271(32):19009-19012.
74. Cho H, Kim TK, Mancebo H, Lane WS, Flores O et al. A protein phosphatase functions to recycle RNA polymerase II. Genes Dev 1999; 13(12):1540-1552.
75. Poon RYC, Yamashita K, Howell M, Ershler MA, Belyavsky A et al. Cell cycle regulation of the $p34^{cdc2}/p33^{cdk2}$-activating kinase $p40^{MO15}$. J Cell Sci 1994; 107:2789-2799.
76. Tassan J-P, Schultz SJ, Bartek J, Nigg EA. Cell cycle analysis of the activity, subcellular localization and subunit composition of human CAK (CDK-activating kinase). J Cell Biol 1994; 127:467-478.
77. Yankulov KY, Bentley DL. Regulation of CDK7 substrate specificity by MAT1 and TFIIH. EMBO J 1997; 16:1638-1646.
78. Lu H, Fisher RP, Bailey P, Levine AJ. The CDK7-cycH-p36 complex of transcription factor IIH phosphorylates p53, enhancing its DNA-binding activity in vitro. Mol Cell Biol 1997; 17:5923-5934.
79. Inamoto S, Segil N, Pan Z-Q, Kimura M, Roeder RG. The cyclin-dependent kinase-activating kinase (CAK) assembly factor, MAT1, targets and enhances CAK activity on the POU domains of octamer transcription factors. J Biol Chem 1997; 272:29852-29858.
80. Ko LJ, Shieh SY, Jayaraman L, Tamai K, Taya Y et al. p53 is phosphorylated by CDK7-cyclin H in a $p36^{MAT1}$-dependent manner. Mol Cell Biol 1997; 17:7220-7229.
81. Martinez A-M, Afshar M, Martin F, Cavadore J-C, Labbé J-C et al. Dual phosphorylation of the T-loop in cdk7: its role in controlling cyclin H binding and CAK activity. EMBO J 1997; 16:343-354.
82. Garrett S, Barton BA, Knights R, Jin P, Morgan DO et al. Reciprocal activation by cyclin-dependent kinases 2 and 7 is directed by substrate specificity determinants outside the T-loop. Mol Cell Biol 2001; 21:88-99.
83. Morgan DO. Cyclin-dependent kinases: engines, clocks and microprocessors. Annu Rev Cell Dev Biol 1997; 13:261-291.
84. Nagahara H, Ezhevsky SA, Vocero-Akbani AM, Kaldis P, Solomon MJ et al. Transforming growth factor beta targeted inactivation of cyclin E:cyclin-dependent kinase 2 (Cdk2) complexes by inhibition of Cdk2 activating kinase activity. Proc Natl Acad Sci USA 1999; 96(26):14961-14966.
85. Murray A, Marks D. Can sequencing shed light on cell cycling? Nature 2001; 409:844-846.
86. Kaldis P, Solomon MJ. Analysis of CAK activities from human cells. Eur J Biochem 2000; 267(13):4213-4221.
87. Fisher RP, Morgan DO. CAK in TFIIH: crucial connection or confounding coincidence? Biochim Biophys Acta 1996; 1288:O7-O10.
88. Nigg EA. Cyclin-dependent kinase 7: At the cross-roads of transcription, DNA repair and cell cycle control? Curr Opin Cell Biol 1996; 8:312-317.

89. Foe VE, Odell GM, Edgar BA. Mitosis and morphogenesis in the *Drosophila* embryo: Point and counterpoint. In: Bate M, Martinez-Arias A, eds. The Development of *Drosophila melanogaster*. CSHL Press, 1993:149-300.
90. Lee D, Lis JT. Transcriptional activation independent of TFIIH kinase and the RNA polymerase II mediator in vivo. Nature 1998; 393:389-392.
91. Espinoza FHE, Farrell A, Nourse JL, Chamberlin HM, Gileadi O et al. Cak1 is required for Kin28 phosphorylation and activation *in vivo*. Mol Cell Biol 1998; 18:6365-6373.
92. Kimmelman J, Kaldis P, Hengartner CJ, Laff GM, Koh SS et al. Activating phosphorylation of the Kin28p subunit of yeast TFIIH by Cak1p. Mol Cell Biol 1999; 19:4774-4787.
93. Larochelle S, Chen J, Knights R, Pandur J, Morcillo P, Erdjument-Bromage H et al. T-loop phosphorylation stabilizes the CDK7-cyclin H-MAT1 complex in vivo and regulates its CTD kinase activity. EMBO J 2001; 20:3749-3759.

Chapter 5

CDK-Activating Kinases in Higher Plants

Masaaki Umeda

Introduction

Recent development in plant molecular genetics has facilitated our understanding of how plants accomplish their body plans in response to various environmental signals. Plants have a unique feature of making almost all organs after embryogenesis, and morphogenesis usually takes place at a specific tissue, called meristems. Several genes that control meristem activities have been characterized by using *Arabidopsis* genetics,[1] but the mechanisms by which such factors cross-talk with cell cycle regulators to balance cell division against cell differentiation in the meristems remain to be elucidated. To determine these mechanisms, studies have first characterized various genes involved in cell cycle regulation in many plant species, such as Alfalfa, *Antirrhinum*, *Arabidopsis*, rice and tobacco.[2-5] Recent reports showed that reverse-genetic approaches with transgenic plants are useful for dissection of the signaling cascade involved in the control of cell division machinery.[6-8]

Here I review CDKs and CDK-activating kinases (CAKs) in plants, and focus on the unique features of rice and *Arabidopsis* CAKs. Since CAK is an upstream kinase which activates almost all CDKs, up- or down-regulation of the kinase activity should critically influence total CDK activity in each cell of the meristems. Therefore, I shall also discuss how CAK may perceive internal and/or external signals and function in the maintenance of meristem activities.

Plant CDKs and Cyclins

As in animals, plants also express several types of CDK, and multiple genes for CDKs have been found in *Arabidopsis*,[9,10] alfalfa,[11-13] rice,[14-16] soybean,[17] maize,[18] and *Antirrhinum*.[19,20] This suggests that the division of plant cells is controlled by specific CDK/cyclin pairs, rather than by a single major CDK as in the case of yeast. However, to date, only little information is available on the active CDK/cyclin complexes in plants.

Plant CDKs can be classified into five groups (A- to E-type) based on amino acid sequence similarities.[21] A consensus has recently emerged indicating that every A-type CDK has the conserved PSTAIRE sequence, and shows a constitutive expression throughout the cell cycle, indicating that they function by interacting with different cyclins in each phase of the cell cycle. In contrast, B-type CDKs are distinct from other CDKs because their transcripts accumulate at a particular phase of the cell cycle, namely during the G2-to-M phase or during the S-to-M phase.[13,20,22,23] Therefore, these CDKs would be activated by interacting with mitotic cyclins. C-type CDKs are similar to the CHED kinase and CDK9 of human, while E-type CDKs are unique to plants and their function remains unknown. D-type CDKs are CDK-activating kinases (CAKs) that will be discussed later.

The CDK-Activating Kinase (CAK), edited by Philipp Kaldis.

A- and B-type cyclins are mitotic cyclins required for progression from the G2 to M phase. Many cDNAs encoding mitotic cyclins have been isolated from various plant species.[24] Based on the amino acid similarity, A- and B-type cyclins have been further classified into small groups, namely CycA1-A3 and CycB1-B2.[24] The growing list of plant mitotic cyclins suggests that the progression from the G2 to M phase is controlled by A- and B-type CDKs in association with several types of cyclins. We have recently shown that transcripts of a rice CycB2 (CycB2;2) starts to accumulate from the G2 phase, but the expression level is reduced during progression of the anaphase.[25] CycB2;2 specifically binds to one of the B-type CDKs in vitro (M. Umeda & H. Uchimiya, unpublished results), suggesting that CycB2 may play an important role in the control of mitotic events in plant cells.

D-type cyclins are G1 cyclins that show cell cycle-independent expression. Plant D-type cyclins are classified into three groups (D1-D3), among which genes encoding CycD3 have been widely identified.[24] Transcripts of *Arabidopsis* CycD2 and CycD3 were induced by the addition of sucrose, and the CycD3 expression was enhanced by phytohormone (cytokinin) application.[26,27] Such induction is similar to animal counterparts whose transcription is also induced by the presence of mitogens. In contrast, the transcripts of two tobacco cyclins (CycD2;1 and CycD3;1) accumulate during mitosis, which is unusual for D-type cyclins.[28] In *Arabidopsis* and tobacco, A-type CDKs are known to interact with CycD.[2,29] Considering that *Arabidopsis* has no counterpart of human CDK4 or CDK6, it is likely that CycD may function in the G1 phase by binding to A-type CDKs in plants.

Vertebrate-Type CAKs in Plants

Primary Structures

The rice cDNA named *R2* was first isolated by Hata[14] as a novel Cdc2/Cdc28-related kinase. Since cDNAs for the subunits of CDK-activating kinase (CAK) have been identified in animals, it is now clear that R2 encodes a homolog of CDK7/p40^{MO15}. CAKs of starfish, *Xenopus* and mammals form a heterotrimeric complex composed of a catalytic kinase subunit CDK7/p40^{MO15},[30-32] a regulatory cyclin H subunit,[33-35] and an assembly factor, MAT1.[36-38] A related CAK complex in *Schizosaccharomyces pombe* consists of Crk1/Mop1/Mcs6 catalytic and Mcs2 regulatory subunit.[39,40] Rice R2 is closely related to CDK7/p40^{MO15} in animal and fission yeast with 55% and 50% homology, respectively, at the amino acid level (Fig. 5.1). Recently, we isolated two *Arabidopsis* homologs, named *cak2At* and *cak3At*, and found that they also showed high homology to vertebrate-type CAKs (Fig. 5.1).

The most striking feature of plant homologs is that they have an extended carboxy-terminal region of 60-70 amino acids. There exists a significant similarity in this region among rice and *Arabidopsis* proteins (Fig. 5.1), suggesting a possible role specific to plants. Overexpression of *R2* or *cak2At* in fission yeast wild-type cells was associated with growth inhibition (M. Umeda & H. Uchimiya, unpublished results).[41] In contrast, overexpression of CDK7/p40^{MO15} of *Xenopus* did not repress cell growth, rather partially overcame the temperature sensitivity of a *crk1/mop1/Mcs6*-deficient mutant.[39] Since deletion of the carboxy-terminal region of R2 resulted in normal growth of wild-type cells, it is likely that the extended carboxy-terminal region is inhibitory on yeast cell growth.

Functional Similarity to Vertebrate-Type CAKs

Cyclin H homologs from poplar and rice plants show approximately 40% amino acid similarity to human cyclin H and fission yeast Mcs2.[42] Yeast two-hybrid and in vitro pull-down assay have demonstrated that rice cyclin H specifically binds to R2 but not to other rice CDKs.[42] Interestingly, R2 protein binds not only to cyclin H but also to cyclin C,[42] whereas human cyclin C interacts with CDK8[43] but not with CDK7/p40^{MO15} in the yeast two-hybrid sys-

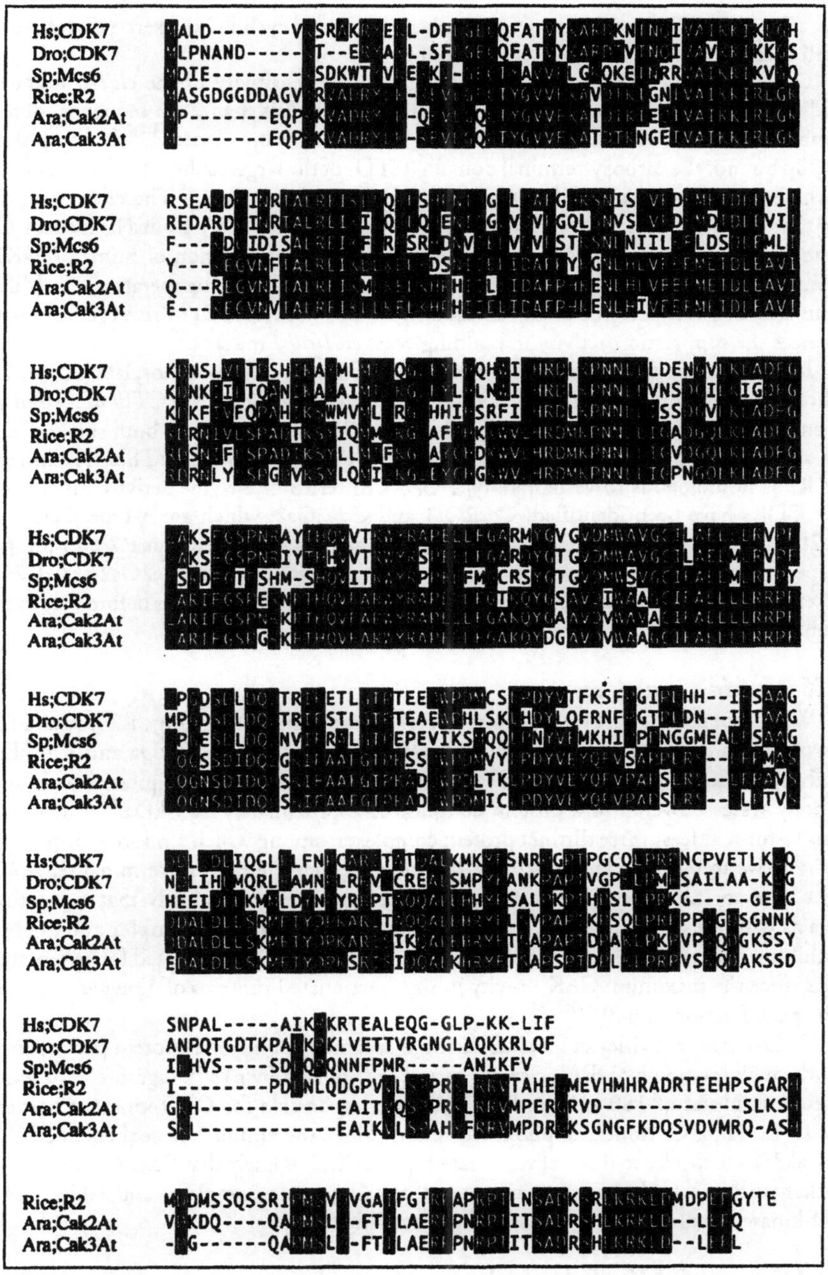

Fig. 5.1. Similarity of amino acid of CDK7/p40^{MO15} family proteins. Amino acid sequences were aligned to yield the highest similarity to each other. Amino acids identical among more than three sequences are shown in white letters. Human (Hs) and *Drosophila* (Dro) CDK7, *Sch. pombe* (Sp) Mcs6, rice R2, and *Arabidopsis* (Ara) Cak2At and Cak3At, are included. The nucleotide sequence data of *cak2At* and *cak3At* will appear in the DDBJ/EMBL/GenBank nucleotide sequence databases under the Accession Nos. AB047274 and AB047275, respectively.

tem[35]. Therefore, R2 may interact with both cyclin H and cyclin C to exert a diverse function in cell division and transcription as discussed below.

Overexpression of *R2* suppresses the temperature sensitivity of the *civ1-4* mutation in budding yeast.[41] Civ1p/Cak1p is a CAK which displays less than 25% sequence identity to members of the CDK family (Fig. 5.2), and is active as a monomer.[44-46] It phosphorylates Cdc28p but not the carboxy-terminal domain (CTD) of the largest subunit of RNA polymerase II, which is another substrate of vertebrate and fission yeast CAKs.[47-49] Therefore, it is remarkable that R2 is capable of overcoming the mutation in *civ1/cak1*. By itself, human CDK7 cannot complement the *civ1/cak1*-deficient mutant, but coexpression of human CDK7 and cyclin H allows slow growth of the *civ1-4* strain at the restrictive temperature (C. Miled, C. Mann and G. Faye, personal communication). In fact, coexpression of rice cyclin H with *R2* enhances the suppression activity in budding yeast *civ1/cak1* mutant.[42]

Immunoprecipitates of rice proteins with anti-R2 antibody phosphorylate human CDK2 at threonine 160 within the T-loop in vitro.[41] They also phosphorylate CTD of *Arabidopsis*.[41] Recombinant R2 protein fused to GST exhibits low kinase activities on both substrates, however, association with cyclin H, significantly elevates the kinase activity.[42] These results indicate that R2 is homologous to vertebrate-type CAKs in terms of enzyme activity. In rice plants, three CDKs have been identified; Cdc2Os1 and Cdc2Os2, which are A-type CDKs,[15] and Cdc2Os3, which is a B-type CDK.[16,23] In vitro kinase assay showed that R2 phosphorylated the specific threonine residue within the T-loop of Cdc2Os1 but not Cdc2Os2 or Cdc2Os3.[41] Therefore, Cdc2Os2 and Cdc2Os3 may require their own cyclin partners before they can serve as substrates for R2, as in the case of human CDC2/cyclin B.[33,50]

CAK Protein Complexes

When rice proteins are fractionated by gel-filtration chromatography, R2 is found mainly in two protein complexes with molecular masses of approximately 70 kDa and 190 kDa, respectively.[41] However, CDK- and CTD-kinase activities immunoprecipitated with anti-R2 antibody were recovered in a protein complex of approximately 105 kDa.[41] Therefore, R2 seems to form at least three distinct protein complexes, among which a minor complex of 105 kDa is active in terms of kinase activity. Considering that the carboxy-terminal region of R2 might have an inhibitory effect on the growth of yeast cells, it is likely that a protein that interacts with the carboxy-terminal region is present in the 190 kDa complex and might have an inhibitory effect on the activity of R2. Such protein has not been proposed for vertebrate-type CAKs since the maximum CAK activity in size-fractionated proteins of *Xenopus* was detected in the peak fraction of $p40^{MO15}$.[31]

CTD-kinase activities of *Arabidopsis* were separated from total protein extract into the flow-through fraction by DEAE-Sepharose and precipitated by $p13^{suc1}$-agarose.[51] At least two protein complexes of 180 kDa and 850 kDa are responsible for CTD-kinase activities (M. Umeda, L. Bakó, C. Koncz, unpublished result). Since the amino acid sequences of Cak2At and Cak3At are similar to those of vertebrate-type CAKs, it is likely that Cak2At and/or Cak3At together with a cyclin H homolog are present in such protein complexes and exhibit not only CTD-kinase but also CDK-activating kinase activities.

The Distinct CAK of *Arabidopsis*

Unique Features of Arabidopsis Cak1At

The *Arabidopsis* cDNA, named *cak1At*, suppresses the *civ1-4* mutation in budding yeast.[51] In contrast to rice R2, overexpression of *cak1At* suppresses the temperature sensitivity of *crk1/mop1/Mcs6* mutation in fission yeast.[51] A predicted protein of Cak1At shows a close similarity to CDK7/$p40^{MO15}$, but the homology is restricted to the conserved kinase domains (Fig. 5.2).

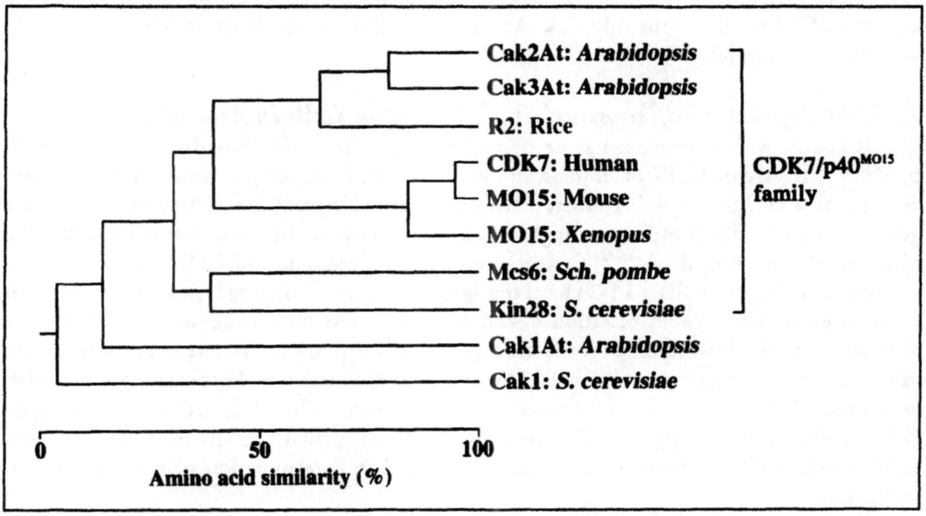

Fig. 5.2. Phylogenetic tree obtained by amino acid sequence comparison of CAK proteins and budding yeast Kin28.

A unique feature is that the Cak1At protein carries an extra stretch of 112 amino acids between the kinase active site and phosphoregulatory site. Although it formed a protein complex of approximately 180 kDa in *Arabidopsis*, rice cyclin H could not interact with Cak1At in yeast two-hybrid assay (M. Umeda, H. Uchimiya, unpublished result). These results suggest that Cak1At may have a distinct function from vertebrate-type CAKs.

Cak1At phosphorylates the specific threonine within the T-loop of human CDK2, which is then activated in terms of histone H1-kinase activity.[51] The tobacco CDK/cyclin D complex is also activated when incubated with recombinant Cak1At expressed in insect cells (H. Nakagami, M. Sekine, personal communication). However, no CTD-kinase activity associated with Cak1At has been observed, and Cak1At was biochemically separated from CTD-kinase activities of *Arabidopsis* protein extract.[51] These data indicate that Cak1At has a CDK-activating kinase activity, but no CTD-kinase activity. Such enzymatic characteristics is similar to that of budding yeast Civ1p/Cak1p, although no significant similarity has been found between Cak1At and Civ1p/Cak1p.

Since Cak1At could interact with Cak2At in yeast two-hybrid assay and in vitro pull-down assay (M. Umeda, H. Uchimiya, unpublished result), it is probable that a functional relationship exists between these different types of CAKs. In fission yeast, *Csk1* has been isolated as an efficient suppressor of the *Mcs2-75* allele.[52] Recent reports showed that Csk1 kinase is an in vivo activating kinase of the Mcs6-Mcs2 CAK[53] and also activates Cdc2 in complexes with either Cdc13 or Cig2 cyclins.[54] Preliminary data from our laboratory suggest that *cak1At* is able to complement *Mcs2* mutant as well as *Csk1/Mcs6* double-mutant of fission yeast (M. Umeda and H. Uchimiya, unpublished results), an indication that Cak1At might function as a CAK-activating kinase (CAKAK) as well as a CAK in *Arabidopsis*. In budding yeast, Civ1p/Cak1p is required for activating phosphorylation of Kin28p as well as that of Cdc28p.[55,56] Kin28p of budding yeast is the closest structural relative of CDK7/p40^{MO15} (Fig. 5.2), and it associates with TFIIH and readily phosphorylates the CTD of RNA polymerase II, while no CAK activity has been demonstrated.[57,58] Since *cak1At* overexpression strongly suppresses the *civ1/cak1* mutation in budding yeast,[51] it is likely that Cak1At may phosphorylate and activate a CTD-kinase, possibly Cak2At and/or Cak3At, in *Arabidopsis*. We are currently verifying this model.

Cak1At Regulates Differentiation of Root Stem Cells in Arabidopsis

Recently, we overexpressed sense or antisense gene for *cak1At* in *Arabidopsis*.[63] When $p34^{cdc2aAt}$, an A-type CDK of *Arabidopsis*, was recovered from the protein extract a few days after germination, its histone H1-kinase activity was reduced in parallel with the sense or antisense gene expression. We found that a higher Cak1At activity in the sense plants increased the phosphorylated form of $p34^{cdc2aAt}$, which resulted in a reduction of CDK activities by yet unknown mechanisms. After 14-24 hr of transgene expression, initial cells in the root meristem were differentiated without cessation of cell division. These results suggest that a moderate expression of *cak1At* is required for the maintenance of the pluripotent state of root initial cells as well as for keeping sufficient level of CDK activity in the meristem. In contrast, when *cak1At* was overexpressed at later developmental stage, no reduction in CDK activity was observed (M. Umeda, H. Uchimiya, unpublished results), suggesting that Cak1At may not be crucial for controlling CDK activities but rather be involved in determination of stem cell fate in meristems.

Regulation of CAK Activities in Plants

Transcriptional Control of Genes for Plant CAKs

In rice plants, R2 and cyclin H are expressed in almost all tissues, but are highly expressed in actively-dividing cells.[42] Transcripts of cyclin H are abundant in the meristematic region of the growing rice internode, but are found also at lower levels in elongating and in differentiated cells in deepwater rice.[42] This is in contrast to mitotic cyclin genes whose expression is restricted to dividing cells.[25] This supports the notion that cyclin H is involved not only in cell division but also in transcription as discussed above.

In partially synchronized suspension cells from rice, R2 mRNA level is elevated in G1 and S phase,[59] and a close correlation has been observed between changes in the S phase population and changes in cyclin H expression.[42] Moreover, the kinetic pattern of cyclin H expression is similar to the population of S-phase cells in the intercalary meristem of deepwater rice after submergence.[42] On the other hand, induction of R2 and cyclin H mRNA in the S phase was not as clear as those of B-type CDK or mitotic cyclins, based on in situ hybridization patterns of mRNAs with root sections.[23,25] Therefore, CAK activity is required throughout the cell cycle, but a higher activity is probably involved in S phase progression of rice cells.

Hormonal Regulation of CAK Activities

In deepwater rice, gibberellin, a phytohormone, promotes histone H1-kinase activity in the intercalary meristem and induces rapid growth of internodes.[60] Expression analysis showed that gibberellin enhanced transcript levels of A- and B-type CDKs, mitotic cyclins (CycB2), and R2.[59,60] This suggests that a signal triggered by gibberellin might be transmitted to a transcription factor that regulates coordinate expression of genes for CDKs, cyclins and CAK.

Transgenic tobacco plants expressing *R2* and cyclin H of rice showed abnormal morphology of leaf, flower and pollen, and partial loss of apical dominance (M. Yamaguchi, M. Umeda, H. Uchimiya, unpublished results). Leaf explants of transgenic lines produced calli on a medium containing auxin but no cytokinin. Since both phytohormones are required for initiation of cell division in this system, cells expressing *R2* and cyclin H might start cell division independent of a trigger with cytokinin. This preliminary result suggests that CAK activity may be regulated downstream of cytokinin signaling pathway. A similar result has been reported for

Arabidopsis CycD3; constitutive expression of *CycD3* in transgenic plants allowed induction and maintenance of cell division in the absence of exogenous cytokinin.[61] Gene expression of *CycD3* was up-regulated with cytokinin,[61,62] thus transcriptional control of *CycD3* would be another target of cytokinin action. These results indicate that cytokinin may play a crucial role in activation of cell division by controlling CDK activities through CAK and D-type cyclin.

Summary and Conclusions

Rice R2 is closely related to CAKs of vertebrate and fission yeast. R2 binds to cyclin H and is activated in terms of CDK- and CTD-kinase activities. R2 forms at least three protein complexes in vivo, namely 70, 105 and 190 kDa complexes, among which the 105 kDa complex has both CDK- and CTD-kinase activities. *Arabidopsis* counterparts have been recently identified as Cak2At and Cak3At that are highly homologous to each other. We are currently testing their enzymatic activities to see whether they are functional homologs of vertebrate-type CAKs.

Arabidopsis Cak1At is a distinct type of CAK, compared with other organisms. The amino acid similarity is restricted to several kinase domains, and Cak1At has CDK-kinase activity but no CTD-kinase activity. Since cyclin H does not interact with Cak1At, it is interesting to identify the regulatory subunit that forms the active complex of 180 kDa with Cak1At. Our genetic data with yeast mutants suggest that Cak1At may control not only CDKs but also Cak2At and Cak3At. This conclusion is surprising because no close relative of Cak1At has been identified in other organisms, and an upstream kinase that regulates CAK activities has been reported only in fission yeast. An interesting question is how the CDK-phosphorylation cascade is controlled by external and internal signals in plants. Cak1At may be also involved in determination of stem cell fate in meristems. Therefore, it is intriguing to see whether there exits any link between genes that regulate meristem function and CAK-mediated CDK-phosphorylation cascade in *Arabidopsis*.

Plant hormones such as auxin and cytokinin are essential for activation of cell division. However, only a limited evidence has been presented regarding the mechanisms by which phytohormones control the cell cycle machinery. CAK may be one of the targets that receive internal signals triggered by cytokinin. Then, the question is how CAK is regulated by cytokinin signal; for example, at transcriptional level or by protein phosphorylation? To answer these questions, transgenic techniques and *Arabidopsis* mutants with defects in CAK genes should be useful for dissecting the signaling cascade. Studies employing such biological tools should enhance our understanding of the regulatory mechanisms underlying the totipotency of plant cells.

Acknowledgments

I am grateful to Prof. Hirofumi Uchimiya (The University of Tokyo) and Dr. Csaba Koncz (Max-Planck-Institut für Züchtungsforschung) for their helpful discussions. Part of this work was supported by a Grant-in-Aid for Scientific Research from the Ministry of Education, Science and Culture of Japan, and by a grant from the Ministry of Agriculture, Forestry and Fisheries of Japan.

References

1. Fletcher JC, Meyerowitz EM. Cell signaling within the shoot meristem. Curr Opin Plant Biol 2000; 3:23-30.
2. Huntley RP, Murray JAH. The plant cell cycle. Curr Opin Plant Biol 1999; 2:440-446.
3. Mironov V, De Veylder L, Van Montagu M et al. Cyclin-dependent kinases and cell division in plants—The nexus. Plant Cell 1999; 11:509-521.
4. den Boer BGW, Murray JAH. Triggering the cell cycle in plants. Trends Cell Biol 2000; 10:245-250.

5. Umeda M. Cyclin-dependent protein kinases and their regulators in plants. Plant Biotech 2000; 17:177-186.
6. Doerner P, Jørgensen J-E, You R et al. Control of root growth and development by cyclin expression. Nature 1996; 380:520-523.
7. Riou-Khamlichi C, Huntley R, Jacqmard A et al. Cytokinin activation of *Arabidopsis* cell division through a D-type cyclin. Science 1999; 283:1541-1544.
8. Cockcroft CE, den Boer BGW, Healy JMS et al. Cyclin D control of growth rate in plants. Nature 2000; 405:575-579.
9. Ferreira PCG, Hemerly AS, Villarroel R et al. The *Arabidopsis* functional homolog of the p34^{cdc2} protein kinase. Plant Cell 1991; 3:531-540.
10. Hirayama T, Imajuku Y, Anai T et al. Identification of two cell-cycle-controlling *cdc2* gene homologs in *Arabidopsis thaliana*. Gene 1991; 105:159-165.
11. Hirt H, Páy A, Györgyey J et al. Complementation of a yeast cell cycle mutant by an alfalfa cDNA encoding a protein kinase homologous to p34^{cdc2}. Proc Natl Acad Sci USA 1991; 88:1636-1640.
12. Hirt H, Páy A, Bögre L et al. *cdc2MsB*, a cognate *cdc2* gene from alfalfa, complements the G1/S but not the G2/M transition of budding yeast *cdc28* mutants. Plant J 1993; 4:61-69.
13. Magyar Z, Mészáros T, Miskolczi P et al. Cell cycle phase specificity of putative cyclin-dependent kinase variants in synchronized alfalfa cells. Plant Cell 1997; 9:223-235.
14. Hata S. cDNA cloning of a novel cdc2$^+$/CDC28-related protein kinase from rice. FEBS Lett 1991; 279:149-152.
15. Hashimoto J, Hirabayashi T, Hayano Y et al. Isolation and characterization of cDNA clones encoding *cdc2* homologues from *Oryza sativa*: A functional homologue and cognate variants. Mol Gen Genet 1992; 233:10-16.
16. Kidou S, Umeda M, Uchimiya H. Nucleotide sequence of rice (*Oryza sativa* L.) cDNA homologous to *cdc2* gene. DNA Sequence 1994; 5:125-129.
17. Miao G-H, Hong Z, Verma DPS. Two functional soybean genes encoding p34^{cdc2} protein kinases are regulated by different plant developmental pathways. Proc Natl Acad Sci USA 1993; 90:943-947.
18. Colasanti J, Tyers M, Sundaresan V. Isolation and characterization of cDNA clones encoding a functional p34^{cdc2} homologue from *Zea mays*. Proc Natl Acad Sci USA 1991; 88:3377-3381.
19. Fobert PR, Coen ES, Murphy GJP et al. Patterns of cell division revealed by transcriptional regulation of genes during the cell cycle in plants. EMBO J 1994; 13:616-624.
20. Fobert PR, Gaudin V, Lunness P et al. Distinct classes of *cdc2*-related genes are differentially expressed during the cell division cycle in plants. Plant Cell 1996; 8:1465-1476.
21. Joubès J, Chevalier C, Dudits D et al. CDK-related protein kinases in plants. Plant Mol Biol 2000; 43:607-621.
22. Segers G, Gadisseur I, Bergounioux C et al. The *Arabidopsis* cyclin-dependent kinase gene *cdc2bAt* is preferentially expressed during S and G2 phases of the cell cycle. Plant J 1996; 10:601-612.
23. Umeda M, Umeda-Hara C, Yamaguchi M et al. Differential expression of genes for cyclin-dependent protein kinases in rice plants. Plant Physiol 1999; 119:31-40.
24. Renaudin JP, Doonan JH, Freeman D et al. Plant cyclins: A unified nomenclature for plant A-, B- and D-type cyclins based on sequence organization. Plant Mol Biol 1996; 32:1003-1018.
25. Umeda M, Iwamoto N, Umeda-Hara C et al. Molecular characterization of mitotic cyclins in rice plants. Mol Gen Genet 1999; 262:230-238.
26. Soni R, Carmichael JP, Shah ZH et al. A family of cyclin D homologs from plants differentially controlled by growth regulators and containing the conserved retinoblastoma protein interaction motif. Plant Cell 1995; 7:85-103.
27. Riou-Khamlichi C, Menges M, Healy JMS et al. Sugar control of the plant cell cycle: Differential regulation of *Arabidopsis* D-type cyclin gene expression. Mol Cell Biol 2000; 20:4513-4521.
28. Sorrell DA, Combettes B, Chaubet-Gigot N et al. Distinct cyclin D genes show mitotic accumulation or constant levels of transcripts in tobacco bright yellow-2 cells. Plant Physiol 1999; 119:343-351.

29. Nakagami H, Sekine M, Murakami H et al. Tobacco retinoblastoma-related protein phosphorylated by a distinct cyclin-dependent kinase complex with Cdc2/cyclin D in vitro. Plant J 1999; 18:243-252.
30. Fesquet D, Labbé JC, Derancourt J et al. The MO15 gene encodes the catalytic subunit of a protein kinase that activates cdc2 and other cyclin-dependent kinases (CDKs) through phosphorylation of Thr161 and its homologues. EMBO J 1993; 12:3111-3121.
31. Poon RYC, Yamashita K, Adamczewski JP et al. The cdc2-related protein $p40^{MO15}$ is the catalytic subunit of a protein kinase that can activate $p33^{cdk2}$ and $p34^{cdc2}$. EMBO J 1993; 12:3123-3132.
32. Solomon MJ, Harper JW, Shuttleworth J. CAK, the $p34^{cdc2}$ activating kinase, contains a protein identical or closely related to $p40^{MO15}$. EMBO J 1993; 12:3133-3142.
33. Fisher RP, Morgan DO. A novel cyclin associates with MO15/CDK7 to form the CDK-activating kinase. Cell 1994; 78:713-724.
34. Labbé JC, Martinez AM, Fesquet D et al. $p40^{MO15}$ associates with a p36 subunit and requires both nuclear translocation and Thr176 phosphorylation to generate cdk-activating kinase activity in *Xenopus* oocytes. EMBO J 1994; 13:5155-5164.
35. Mäkelä TP, Tassan JP, Nigg EA et al. A cyclin associated with the CDK-activating kinase MO15. Nature 1994; 371:254-257.
36. Devault A, Martinez AM, Fesquet D et al. MAT1 ('menage à trois') a new RING finger protein subunit stabilizing cyclin H-cdk7 complexes in starfish and *Xenopus* CAK. EMBO J 1995; 14:5027-5036.
37. Fisher RP, Jin P, Chamberlin HM et al. Alternative mechanisms of CAK assembly require an assembly factor or an activating kinase. Cell 1995; 83:47-57.
38. Tassan JP, Jaquenoud M, Fry AM et al. In vitro assembly of a functional human CDK7-cyclin H complex requires MAT1, a novel 36 kDa RING finger protein. EMBO J 1995; 14:5608-5617.
39. Buck V, Russell P, Millar JBA. Identification of a cdk-activating kinase in fission yeast. EMBO J 1995; 14:6173-6183.
40. Damagnez V, Mäkelä TP, Cottarel G et al. *Schizosaccharomyces pombe* Mop1-Mcs2 is related to mammalian CAK. EMBO J 1995; 14:6164-6172.
41. Yamaguchi M, Umeda M, Uchimiya H. A rice homolog of Cdk7/MO15 phosphorylates both cyclin-dependent protein kinases and the carboxy-terminal domain of RNA polymerase II. Plant J 1998; 16:613-619.
42. Yamaguchi M, Fabian T, Sauter M et al. Activation of CDK-activating kinase is dependent on interaction with H-type cyclins in plants. Plant J 2000; 23:1-16.
43. Tassan JP, Jaquenoud M, Leopold P et al. Identification of human cyclin-dependent kinase 8, a putative protein kinase partner for cyclin C. Proc Natl Acad Sci USA 1995; 92:8871-8875.
44. Espinoza FH, Farrell A, Erdjument-Bromage H et al. A cyclin-dependent kinase-activating kinase (CAK) in budding yeast unrelated to vertebrate CAK. Science 1996; 273:1714-1717.
45. Kaldis P, Sutton A, Solomon MJ et al. The Cdk-activating kinase (CAK) from budding yeast. Cell 1996; 86:553-564.
46. Thuret JY, Valay JG, Faye G et al. Civ 1 (CAK in vivo), a novel Cdk-activating kinase. Cell 1996; 86:565-576.
47. Roy R, Adamczewski JP, Seroz T et al. The MO15 cell cycle kinase is associated with the TFIIH transcription-DNA repair factor. Cell 1994; 79:1093-1101.
48. Serizawa H, Mäkelä TP, Conaway JW et al. Association of Cdk-activating kinase subunits with transcription factor TFIIH. Nature 1995; 374:280-282.
49. Shiekhattar R, Mermelstein F, Fisher RP et al. Cdk-activating kinase complex is a component of human transcription factor TFIIH. Nature 1995; 374:283-287.
50. Desai D, Wessling HC, Fisher RP et al. Effects of phosphorylation by CAK on cyclin binding by CDC2 and CDK2. Mol Cell Biol 1995; 15:345-350.
51. Umeda M, Bhalerao RP, Schell J et al. A distinct cyclin-dependent kinase-activating kinase of *Arabidopsis thaliana*. Proc Natl Acad Sci USA 1998; 95:5021-5026.
52. Molz L, Beach D. Characterization of the fission yeast Mcs2 cyclin and its associated protein kinase activity. EMBO J 1993; 12:1723-1732.

53. Hermand D, Pihlak A, Westerling T et al. Fission yeast Csk1 is a CAK-activating kinase (CAKAK). EMBO J 1998; 17:7230-7238.
54. Lee KM, Saiz JE, Barton WA et al. Cdc2 activation in fission yeast depends on Mcs6 and Csk1, two partially redundant Cdk-activating kinases (CAKs). Curr Biol 1999; 9:441-444.
55. Espinoza FH, Farrell A, Nourse JL et al. Cak1 is required for Kin28 phosphorylation and activation in vivo. Mol Cell Biol 1998; 18:6365-6373.
56. Kimmelman J, Kaldis P, Hengartner CJ et al. Activating phosphorylation of the Kin28p subunit of yeast TFIIH by Cak1p. Mol Cell Biol 1999; 19:4774-4787.
57. Feaver WJ, Svejstrup JQ, Henry NL et al. Relationship of CDK-activating kinase and RNA polymerase II CTD kinase TFIIH/TFIIK. Cell 1994; 79:1103-1109.
58. Cismowski MJ, Laff GM, Solomon MJ et al. KIN28 encodes a C-terminal domain kinase that controls mRNA transcription in *Saccharomyces cerevisiae* but lacks cyclin-dependent kinase-activating kinase (CAK) activity. Mol Cell Biol 1995; 15:2983-2992.
59. Sauter M. Differential expression of a CAK (cdc2-activating kinase)-like protein kinase, cyclins and *cdc2* genes from rice during the cell cycle and in response to gibberellin. Plant J 1997; 11:181-190.
60. Sauter M, Mekhedov SL, Kende H. Gibberellin promotes histone H1 kinase activity and the expression of *cdc2* and cyclin genes during the induction of rapid growth in deepwater rice internodes. Plant J 1995; 7:623-632.
61. Riou-Khamlichi C, Huntley R, Jacqmard A et al. Cytokinin activation of *Arabidopsis* cell division through a D-type cyclin. Science 1999; 283:1541-1544.
62. Soni R, Carmichael JP, Shah ZH et al. A family of cyclin D homologs from plants differentially controlled by growth regulators and containing the conserved retinoblastoma protein interaction motif. Plant Cell 1995; 7:85-103.
63. Umeda M, Umeda-Hara C, Uchimiya H. A cyclin-dependent kinase-activating kinase regulates differentiation of root initial cells is *Arabidopsis*. Proc Natl Acad Sci USA 2000; 97:13396-13400.

Chapter 6

Structural Aspects of CDK Activation

Nick R. Brown and Jane A. Endicott

Introduction

Sequential activation of members of the cyclin-dependent kinase (CDK) family orders the events required for DNA replication and cell division. Both the CDK and cyclin families have multiple members that selectively bind to preferred partners to generate complexes that drive cell cycle progression. CDK/cyclin assemblies are regulated by multiple mechanisms that include phosphorylation and association with additional regulatory proteins. A number of the structural principles underlying CDK regulation have been elucidated following the determination of a series of CDK and cyclin structures using X-ray crystallography and NMR methods.

CDK Activation

The Structure of Monomeric CDK2

CDK2 encodes little more than the protein kinase catalytic core which is composed of multiple conserved subdomains found in all protein kinases.[1,2] Monomeric, unphosphorylated CDK2 (CDK2) adopts the characteristic protein kinase fold that was first observed following the determination of the structure of cAMP-dependent protein kinase (PKA),[3] (Fig. 6.1a). CDK2 has a largely β sheet N-terminal domain that contains one helix, the C helix, connected via a flexible hinge to a mostly α helical C-terminal domain. The ATP binding site is found at the interface between the two lobes.

A comparison of the structure of CDK2 with that of PKA, a constitutively active protein kinase, shows that the disposition of two key structural elements largely explains monomeric CDK2's inactivity.[1] These elements are the C-helix that contains the PSTAIRE motif (single letter amino acid code, residues 45-51), and the activation segment (residues 145-172 between and including the conserved DFG and APE motifs),[2] (Fig. 6.1A). The CDK2 C-helix is displaced out from the main body of the protein and as a result, the stabilising interaction between Glu51 (the E of the PSTAIRE sequence) and Lys33 is lost. This salt bridge is important for correct alignment of the ATP triphosphate moiety for catalysis, although the inactive monomer can bind ATP.[1] The fold of the CDK2 activation segment both buries the side-chain of Thr160 (the site of activating phosphorylation) away from solvent and precludes substrate binding. It extends round from the back of the ATP binding site to occlude the turn of the conserved glycine-loop that links β1 to β2 from solvent before heading back in an extended conformation into the C-terminal lobe.

The CDK-Activating Kinase (CAK), edited by Philipp Kaldis.

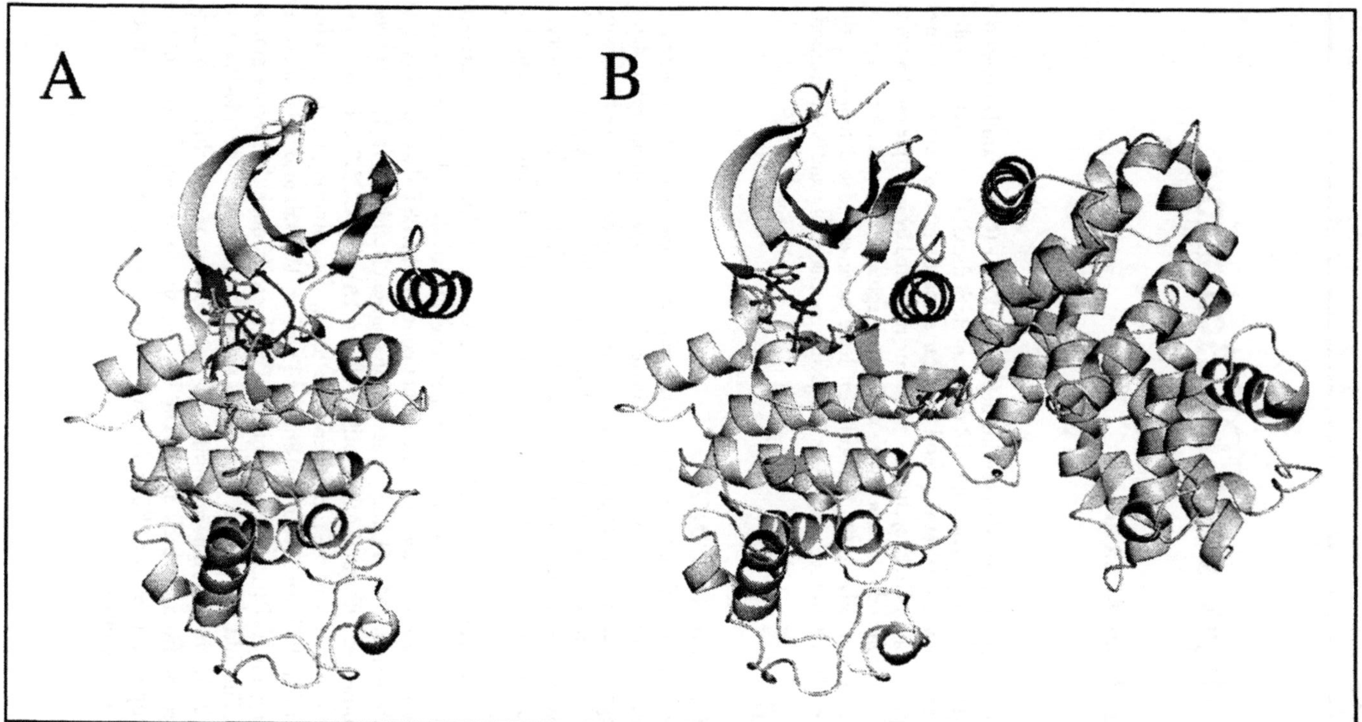

Fig. 6.1. CDK activation. The structures of monomeric CDK2 (PDB code 1hck) (A) and T160pCDK2/cyclin A in complex with a peptide substrate (PDB code 1qmz) (B) were superposed using the CDK2 C-terminal lobe. The CDK2 and cyclin folds are drawn in ribbon representation and colored in gold and in green respectively. The CDK2 glycine loop (residues 11 to 16) is highlighted in pink, the C-helix (residues 45-55) in red and the activation loop (residues 145-172) in cyan. AMP-PNP is drawn in ball and stick representation with carbon atoms in green, nitrogens in blue and oxygen atoms coloured red. The phospho-Thr160 side chain is shown. The figure highlights the large structural rearrangements within CDK2, particularly of the C-helix and of the activation loop that accompany cyclin binding and phosphorylation of Thr160. A color version of this figure is included in the insert that precedes Chapter 1.

CDK2 Activation by Cyclin Binding

The unphosphorylated CDK2/cyclin A (T160CDK2/cyclin A) structure revealed that when cyclin A binds to CDK2, the latter undergoes substantial structural change to create the ATP triphosphate recognition site.[4] The C-helix, which makes numerous contacts with cyclin A, is displaced back into the cleft (Fig. 6.1B). To accommodate this movement a short helix (αL12) located at the N-terminal end of the activation segment melts to form a β-strand, β9 (Fig. 6.1). Residues at the start of β9 that include Asp145, Phe146 and Gly147 (the "DFG" motif) shift so that Asp145 can coordinate to the ATP-associated magnesium ion. The rest of the activation loop moves out of the active site to adopt an extended conformation along the base of the catalytic cleft, with the consequent exposure of the side-chain hydroxyl group of Thr160 to solvent. The overall outcome of CDK2 association with cyclin A is that catalytically important residues necessary for productive MgATP binding and phosphotransfer are appropriately reoriented and the blockade of the active site by the activation segment is relieved. These changes result in an enzyme that exhibits circa 0.2% activity of the fully activated complex.[4] The exposure of Thr160 to solvent has been used as a structural explanation for the observation that cyclin binding promotes Thr160 phosphorylation.

A C-terminal fragment of cyclin A was the target of crystallisation trials for structure determination of both monomeric cyclin A[5] and a CDK2/cyclin A complex.[4] There are no changes in the structure of this fragment on CDK2 binding suggesting that it acts as a rigid template complementary to the active structure of the flexible CDK2 enzyme. The cyclin A structure showed that the cyclin box fold, predicted from sequence analysis to be a protein domain,[6] is a five α-helical bundle. Unexpectedly, given the lack of any sequence similarity, the cyclin A structure also revealed that the cyclin A sequence C-terminal to the cyclin box sequence also adopts a cyclin box fold. The cyclin box fold has since been shown to be a conserved structural element within a number of proteins involved in both cell cycle and transcription control (reviewed in refs. 7 and 8).

The structure of CDK2 in complex with a viral M-type cyclin has since shown that the essential elements of CDK activation promoted by cyclin binding are conserved.[9] In agreement with results reported earlier for T160CDK2/cyclin A, the unphosphorylated CDK2/cyclin M complex shows partial histone H1 kinase activity which can be increased (in this case) circa 50-fold by phosphorylation of CDK2 on Thr160. CDK2 and cyclin M do not have the same relative disposition as CDK2 and cyclin A and, as a result, structurally equivalent cyclin residues make different interactions with CDK2. Presumably this altered interface reflects a slightly different optimal packing arrangement as the flexible CDK2 responds to an alternative template.

The major difference between the two cyclin-bound CDK2 structures is in the activation loop region. Whereas when bound to cyclin A, residues in the CDK2 activation loop following Pro155 adopt a loop structure thrown out towards the solvent,[4] when bound to cyclin M they form a β–strand that forms part of a 3-stranded sheet directly below the C-helix.[9] The two cyclin-bound CDK2 structures continue to differ for the rest of the activation loop sequence. The activation loop β-sheet structure does not extend to Thr160 which appears to be still solvent-exposed and in a loop structure accessible for phosphorylation by CAK. Notably an overlay of the CDK2/cyclin M and T160pCDK2/cyclinA/peptide substrate structures shows that the cyclin M-bound CDK2 activation loop structure adopts a conformation that would clash with a substrate bound to fully active CDK2. As previously observed in the CDK2/cyclin A structure, the adoption by the activation loop of an appropriate structure for substrate recognition has been proposed to require phosphorylation of Thr160.[9]

CDK2 Activation by Phosphorylation

For full catalytic activity CDKs require phosphorylation on a conserved threonine residue (Thr160 in CDK2) that lies within the activation loop. Two forms of CDK-activating kinase (CAK) activity have been discovered. In metazoans it exists as a trimeric complex of CDK7/cyclin H (homologous to Mcs6/Mcs2 in *S. pombe*) and MAT1, whereas in the budding yeast, the conservation of structure and function is lost and a monomeric kinase (termed Cak1p or Civ1p) provides the activity.[10-12] To prepare phosphorylated CDK2/cyclin A complexes for structural studies, both forms of CAK activity have been used. Initially this work involved in vitro phosphorylation of CDK/cyclin complexes using either CDK7/cyclin H prepared from recombinant baculoviral-infected insect cells[13] or Cak1p/Civ1p prepared from *E. coli* cells.[14] Coexpression of CDK2 and Cak1p/Civ1p in *E. coli* cells has been reported[15] and this development has led to a reliable large-scale supply of phosphorylated CDK2 for structural studies.

The Structure of Monomeric Phosphorylated CDK2

Monomeric CDK2 phosphorylated on Thr160 (T160pCDK2), and T160CDK2/cyclin A both exhibit histone H1 kinase activity corresponding to approximately 0.2% of that observed with the fully activated phosphorylated binary complex (T160pCDK2/cyclin A).[14] The structures of T160pCDK2, in the presence and absence of ATP, are similar (rms difference in comparable Cα positions of 0.5 Å) to those of unmodified CDK2 with and without the added ligand.[14,1] The major difference between the four structures occurs in the region of the activation segment. This loop adopts a single conformation in the T160CDK2/ATP complex structure,[14] but is more disordered in the absence of bound ATP.[1] Phosphorylation of Thr160 in the presence or absence of bound ATP further enhances the structural disorder of this part of the molecule with effects extending to atoms that are 13 Å from the site of phosphorylation. The Thr160 hydroxyl oxygen in the T160CDK2 structure is 3.4 Å from a carboxyl oxygen of Glu12 in the glycine-rich ATP binding loop, strongly suggesting that a rearrangement of this region must take place on Thr160 phosphorylation. The structural disorder of the activation loop has been proposed to explain T160pCDK2's low level of kinase activity: At any time a small percentage of the protein may be in a conformation which is compatible with substrate recognition and catalysis.[14]

The Structure of CDK2 Phosphorylated on Thr160 in Complex with Cyclin A

Further structural rearrangements, centred within the activation segment accompany phosphorylation of Thr160 as the CDK2/cyclin A structure converges towards (but remains distinct from) that of active PKA[13] (Fig. 6.2). Considering that phosphorylation of Thr160 leads to an increase of activity of 2 to 3 orders of magnitude, the structural differences are subtle. The phospho-threonine residue acts as an organising centre and a major outcome of the realigned phosphorylated CDK2/cyclin A activation loop structure is the formation of the protein-substrate recognition site. The catalytic and ATP binding residues show little detectable change. The Thr160 phospho group binds neatly into a positively charged pocket formed from Arg50 from the C helix, Arg126 the residue that precedes the catalytic base, and Arg150 at the start of the activation segment. Arg50 and Arg150 also hydrogen bond to the main chain carbonyl oxygens of cyclin A residues Phe267 and Glu268. There are no hydrogen bonds between the phospho-threonine and cyclin A but there are several van der Waals interactions that result in shielding of the phosphothreonine residue by cyclin A from solvent.[13] This result was perhaps not unexpected in the light of earlier studies which had shown that phosphorylation of Thr160 (or of the equivalent residue in CDK1) was of varying importance in determining the stability of CDK-cyclin association.[16] More recently, fluorescence[17] and surface plasmon resonance experiments (N.R.B. unpublished results) with purified proteins have indicated that

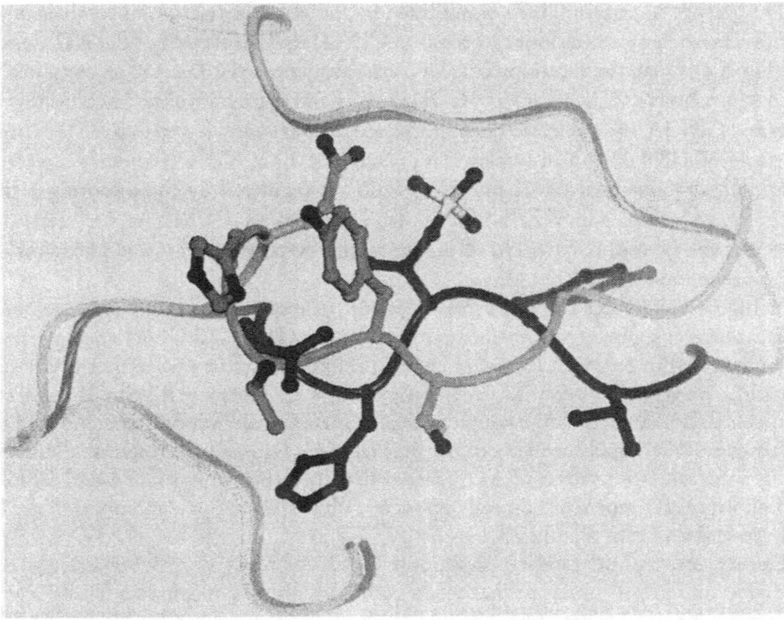

Fig. 6.2. Structural changes in the CDK2 activation loop accompany Thr160 phosphorylation and substrate peptide binding. The structures of T160CDK2/cyclin A (PDB code 1fin) and T160pCDK2/cyclin A in complex with a peptide substrate (PDB code 1qmz) were superposed. The activation loops of the two proteins (residues 145-172) are drawn in worm representation and for clarity, only the side chains of residues 158-162 are included. The T160CDK2/cyclin A structure is colored green and T160pCDK2/cyclin A in complex with a peptide substrate is shown in magenta. Glu162 in the T160CDK2/cyclin A structure occupies an equivalent position to the phosphorylated Thr160 residue in the fully-activated structure (magenta). A color version of this figure is included in the insert that precedes Chapter 1.

there is no significant difference between the affinity of CDK2 for cyclin A and T160pCDK2 for cyclin A. Within the wider protein kinase superfamily, this residue is found to be essential for the structural integrity of the protein kinase fold. Either a phosphorylated serine or threonine or a glutamate residue is conserved at this position in the protein kinase family.[2,18]

The in vivo pathway of assembly of the fully activated T160pCDK2/cyclin A complex has not been determined, but alternatives have been proposed.[19] T160pCDK2 may represent an intermediate on this pathway, and as such could be biologically relevant. The conformation of the activation loop in the T160CDK2/cyclin A structure is dependent in part upon the position of Glu162. Glu162 interacts with the conserved cluster of arginine residues that ultimately forms the binding pocket for phosphorylated-Thr160. CDKs 4 and 6 are also phosphorylated by CDK7/cyclin H.[20] However, both proteins have a valine residue at the position equivalent to Glu162 and do not have an alternative negatively charged amino acid in their activation loop sequences. CDK/cyclin complexes rather than monomeric CDKs have been shown to be the preferred substrates for CDK7/cyclin H in vitro (ref. 20 and N.R.B. unpublished results). These results suggest that either the activation loop sequence may remain flexible subsequent to cyclin binding or that it can adopt alternative conformations in different CDK/cyclin complexes all of which present the threonine hydroxyl as a CDK7/cyclin H substrate.

CDK2 Activation by Dephosphorylation

The activity of certain CDKs is inhibited by the phosphorylation of residues in the conserved glycine-rich motif (residues 11 to 16 in CDK2) that forms part of the ATP-binding site cleft[2] (Fig. 6.1). With the exception of CDK7, the sequences of CDKs1-8 all contain a tyrosine residue equivalent to CDK2 Tyr15.[21] CDKs 1, 2, 4 and 6 are phosphorylated on this tyrosine in vivo and CDKs 1 and 2 are also phosphorylated on Thr14 (reviewed in ref. 21). Thr14 is not conserved in CDK4 or CDK6 (where it is an alanine), or in CDK7 (where it is replaced by a glutamine). The extent of CDK phosphorylation is regulated by the opposing activities of CDC25 phosphatases and WEE1 kinases. In *S. pombe* cells, WEE1 under certain circumstances can also phosphorylate Thr14 but in higher eukaryotes Thr14 is phosphorylated by MYT1, a membrane-associated kinase.

In higher eukaryotes the cell cycle can arrest in response to DNA damage during G1, during S-phase and prior to entry into mitosis [(at the G1, S and G2/M checkpoints respectively), (reviewed in refs. 22, 23, 24)]. These checkpoints act to ensure that ultimately sister chromatid separation during mitosis will only occur with complete fidelity. If the accuracy of this process is decreased, then the daughter cells may accrue mutations that could lead to loss of cell cycle control and uncontrolled cellular proliferation. One cellular response to ultraviolet or ionising radiation is activation of ATM kinase that in turn activates Chk1 and Chk2-dependent pathways that, dependent on cell cycle stage, ultimately regulate members of the CDC25 family (reviewed in refs. 24 and 25).

The structures of the catalytic domains of Cdc25A[26] and Cdc25B[27] revealed that though their shared fold is not similar to that of the protein tyrosine phosphatase family, mechanistically they are identical. These phosphatases all share a conserved catalytic $CysX_5Arg$(Ser/Thr) motif and the dephosphorylation reaction proceeds in a two-step mechanism via a cysteinyl-phosphate enzyme intermediate (reviewed in ref. 28). Unlike other dual-specificity phosphatases which in general show little substrate preference, Cdc25s only dephosphorylate CDKs within the active site. The structure of a Cdc25A substrate, CDK2/cyclin A phosphorylated on Tyr15 and Thr160 has revealed that the phosphotyrosine is exposed to solvent on a flexible loop (J. Tucker, J.A.E., unpublished results). The active site of Cdc25 is very shallow and so it has been suggested that to account for its specificity an extensive part of its surface may be involved in CDK recognition. A precedent for this mechanism has been set by the determination of the structure of CDK2 phosphorylated on Thr160 in complex with kinase-associated phosphatase (KAP).[29] Like Cdc25, KAP is highly selective in its substrate and only dephosphorylates monomeric CDK2 phosphorylated on Thr160. The complex structure shows that the phosphorylated Thr160 residue is at the tip of an exposed loop and that there are few interactions with KAP in this part of the structure. The major protein interface is an extensive one between the KAP C-terminal helix and the CDK2 C-terminal lobe and overlaps in part with the CDK2 CKS binding site.

The flexibility and accessibility of the regions surrounding the Tyr15 and Thr160 residues, both before and after phosphorylation, may have important implications for their recognition by the kinases (WEE1 and CAKs respectively) and phosphatases (Cdc25 and KAP respectively) that modify CDKs at these two sites. The structural conservation of the catalytic sites in active kinase conformations suggests that the protein substrate needs to be able to adopt an approximately extended conformation for recognition.

Structural Insights into CDK Substrate Selection

The Structure of a T160pCDK2/cyclin A/Substrate Peptide Complex

CDKs 1, 2, 4 and 6 phosphorylate the consensus sequence S/T*-P-X-K/R, where S/T* represents the phosphorylated residue and X represents any amino acid. The structure of a

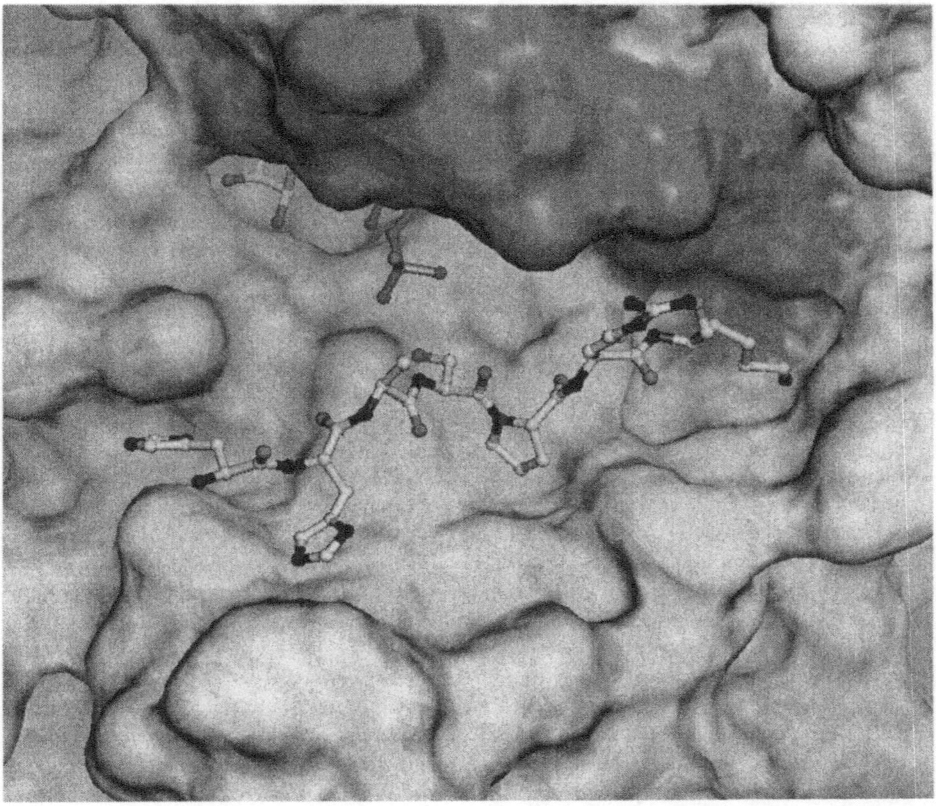

Fig. 6.3. The binding of the consensus substrate peptide to CDK2. The structure of a phosphorylated CDK2/cyclin A/AMPPNP/ Mg^{2+}/peptide complex (PDB code 1qmz) in the vicinity of the catalytic site is shown. The CDK2/cyclin A molecular surface is colored according to the color scheme employed in Figure 6.1: CDK2 in gold, cyclin A in green, and residues from the glycine loop, C-helix and activation loop coloured pink, red and cyan respectively. ATP and the substrate peptide are drawn in ball and stick representation with the carbon atoms in the ATP molecule coloured green and those of the substrate peptide coloured yellow. The CDK2 binding pockets for the substrate amino acid side-chains of histidine (P-2), proline (P+1) and lysine (P+3) are clearly seen. A color version of this figure is included in the insert that precedes Chapter 1.

T160pCDK2/cyclin A/AMP-PNP/substrate peptide complex provided the first picture of the CDK2 active site with all the essential residues oriented for phospho-transfer.[15] The substrate peptide (HHASPRK) binds in an extended conformation across the kinase surface primarily contacting the activation segment (Fig. 6.3).

The substrate residues N-terminal to the phosphorylatable serine residue (i.e., P-1 to P-3) have few contacts with CDK2. The phosphorylatable serine of the substrate (P0) hydrogen bonds to Asp127 (the catalytic base) and to Lys129 and in this orientation its lone pair of electrons are directed in-line to the βγ-bridging oxygen of AMP-PNP through the γ-phosphorus atom. The catalytic base is thought to assist ionisation of the serine hydroxyl prior to its nucleophilic attack on the phosphorus atom of the γ–phosphate. Lys129 is proposed to assist catalysis by stabilising the negative charge in the transition state. A single magnesium ion is seen in the structure that chelates the α– and γ–phosphates.

The cavity into which the substrate P+1 proline neatly fits is particularly conspicuous because its creation requires Val164 in the activation segment to adopt an unusual backbone conformation. Val164's main chain carbonyl group points away from the substrate to hydrogen bond with the side chain of Arg169 and its main chain amide interacts with the carbonyl group of Arg126. This valine conformation is only achieved once Thr160 is phosphorylated. In the T160CDK2/cyclin A structure the proline pocket is blocked by the main chain oxygen of Val163.[4] The side chain of Arg169 excludes a large residue at P+1 and the binding of any other residue at this site is disfavoured because of an uncompensated hydrogen bond from the substrate to a main chain nitrogen atom. The proline residue at the P+1 position also restricts the conformation of the rest of the substrate peptide so that the side chain of the P+3 residue points towards the Thr160 phosphate.

The P+2 arginine is directed into the solvent and appears to make few contacts with the enzyme despite the fact that biochemical studies reveal a clear preference for a basic residue at this position.[30] Perhaps long-range charge-charge interactions facilitate the approach of the peptide to the enzyme. The P+3 lysine hydrogen bonds to both the Thr160 phosphate and notably to the main chain oxygen of Ile270 of cyclin A. This is the only contact the substrate makes with the cyclin subunit. It is possible that residues more C-terminal to the phosphorylation site in a longer peptide or protein substrate would make contact with, and whose binding could be influenced by, the cyclin subunit.

Other proline-directed protein kinases, notably members of the ERK (MAP) kinase and glycogen synthase kinase 3 (GSK3) families share a conserved arginine residue equivalent to CDK2 Arg169.[2] This arginine residue is conserved in other metazoan CDK family members, and of the wider CDK family in *S. cerevisiae* in Cdc28p, Kin28p, Pho85p and Srb10p, but not in Cak1p [where it is replaced by a lysine, as in the casein kinase II family[2]]. As both Kin28p and CDK7 are not proline-directed and phosphorylate a threonine residue in the sequence THE, the valine residue equivalent to CDK2 Val164 may be able to adopt a more conventional backbone conformation in these enzymes. The activation loop would then potentially be able to form an anti-parallel β-sheet structure with the substrate as has been observed in other protein kinase/substrate complexes.[31,32]

Role of CDK Binding Proteins in Substrate Selection

Given the simplicity of the CDK substrate recognition sequence around the site of phosphotransfer it was predicted that additional structural and/or sequence motifs would assist CDK substrate selection. The cyclins and CKSs, two protein families that bind to CDKs have been shown to contain such elements and enhance CDK activity towards certain proteins.

Two short consensus sequences one present in a subset of cyclins, the other a cyclin-binding motif present in certain CDK substrates, have been shown to contribute to CDK substrate selection. CDK2/cyclin A can form stable complexes with a number of proteins (for example E2F-1, p107 and p130 and all members of the Cip/Kip CKI family) that contain a ZRXL motif (single letter amino acid code, where Z and X are typically basic).[33,34] Structures of complexes between CDK2/cyclin A/p27[35] and between CDK2/cyclin A/p107 peptide[15] have shown that this motif binds to a hydrophobic surface on cyclin A that is conserved between the A, B, D and E cyclins[5] (Fig. 6.4). This surface lies on the opposite side of the molecule to the CDK2 binding site and it has been proposed to recruit both substrates and inhibitors to cyclin A, B, D and E-dependent kinases by increasing their local concentration.[33] It is not known whether simultaneous binding of substrate to the two sites is a requirement for substrate phosphorylation. However, recruited CDK substrates show a wide variation in the inter-motif distance that argues that there may not be a strict requirement for a particular spatial disposition of the two sequences. Structures of intact CDK substrates that contain both motifs will address this issue.

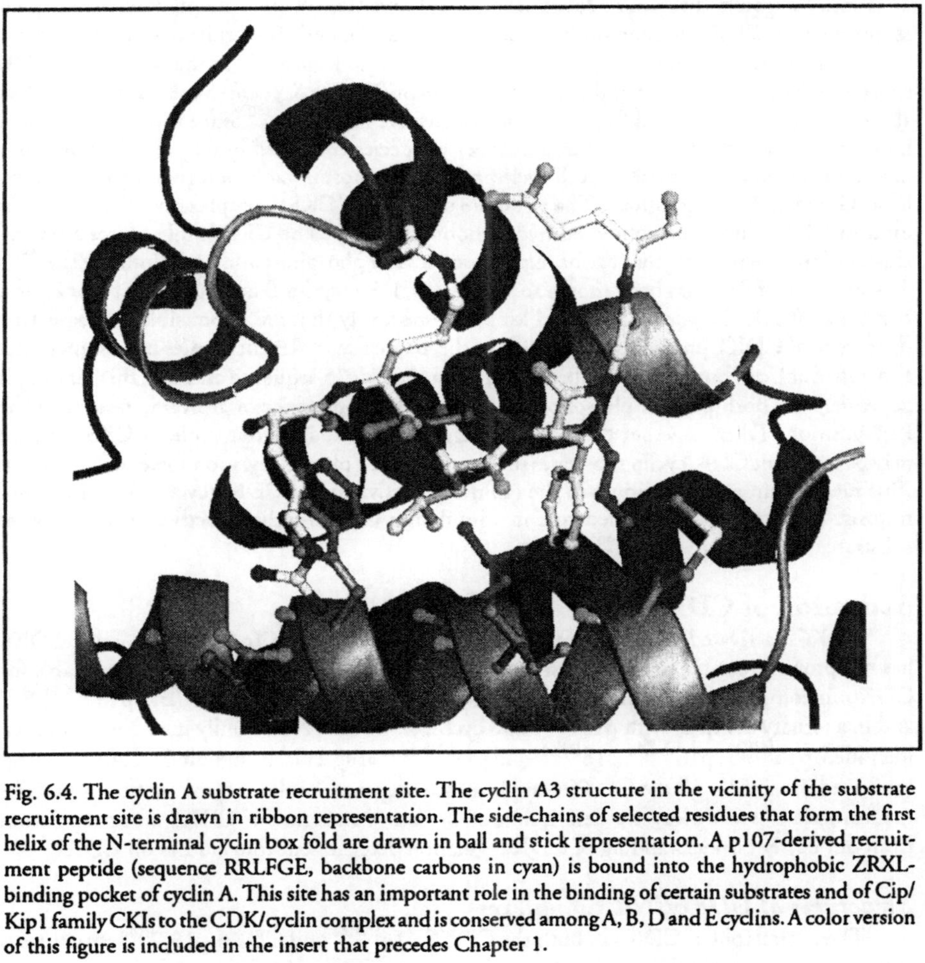

Fig. 6.4. The cyclin A substrate recruitment site. The cyclin A3 structure in the vicinity of the substrate recruitment site is drawn in ribbon representation. The side-chains of selected residues that form the first helix of the N-terminal cyclin box fold are drawn in ball and stick representation. A p107-derived recruitment peptide (sequence RRLFGE, backbone carbons in cyan) is bound into the hydrophobic ZRXL-binding pocket of cyclin A. This site has an important role in the binding of certain substrates and of Cip/Kip1 family CKIs to the CDK/cyclin complex and is conserved among A, B, D and E cyclins. A color version of this figure is included in the insert that precedes Chapter 1.

Members of the Cyclin D family are unique amongst the cyclins in containing an LXCXE motif (where X is any amino acid) N-terminal to the first putative cyclin box fold. This short motif is common to a number of proteins that can associate with the product of the retinoblastoma gene, RB. The structure of the A/B pocket region of pRB in complex with an LXCXE-containing peptide derived from the human papilloma virus pRB-binding protein E7 has provided a model for how pRB might interact with cyclin D via this sequence.[36] The 'A' and 'B' sequences of the pRB pocket encode a tightly associated tandem duplication of the cyclin box fold and the LXCXE motif binds in an extended conformation to the fold encoded by the 'B' pocket.

There are 16 CDK consensus phosphorylation sites in pRB, a subset of which are differentially phosphorylated by CDK4/cyclin D1 and CDK2/cyclin E in vivo (reviewed in ref. 37). There is now evidence to support a model in which pRb phosphorylation by CDK2/cyclin E requires prior phosphorylation by cyclin D-dependent CDKs[38] and that these phosphorylation events are associated with remodelling of pRB-containing complexes and intramolecular structural rearrangments within pRb.[39] The pRB sequence contains a consensus RXL motif.

This result suggests that CDK-dependent phosphorylation of pRB might result from pRB recruitment to CDKs through both LXCXE and RXL-dependent recruitment mechanisms.

Members of the CKS family have also been shown to have a role in the selection of CDK substrates. The first member of this family to be identified was *S. pombe* Suc1 on the basis of its ability to suppress certain CDK1 temperature-sensitive mutations.[40] Since then closely related homologues have been identified in a number of species (reviewed in ref. 41). CKS proteins have multiple roles in controlling cell cycle progression, most notably in regulating CDK activity at G2/M and during mitosis. The structure of a CDK2/CKS1 complex revealed that CKSs bind to CDKs through a conserved hydrophobic surface.[42] The CKS family shares a second cluster of surface residues that has been proposed to be a phospho-amino acid binding site.[43,44] The presence of Suc1 has been shown to stimulate CDK1/cyclin B activity against Cdc25 and Myt1 and Wee1, the phosphatase and kinases respectively that are responsible for regulating the extent of CDK1 phosphorylation within the active site.[45] Recent studies have shown that *S. pombe* Suc1 can bind to a peptide derived from the Cdc25 sequence and that this binding is dependent on both peptide phosphorylation and on the presence of a proline residue at the P+1 position. Taken together these results suggest that one important role of CKS proteins may be to recruit CDK/cyclin substrates to the enzyme in a phosphorylation-dependent manner. This mechanism may contribute to the rapid auto-activation of CDK1/cyclin B at entry into mitosis. Suc1 may have a related role in stimulating CDK1/cyclin B activity towards other substrates later during mitosis.[46,47]

Activation of CDK6

CDK6 is activated by binding members of the cyclin D family. To date, monomeric CDK6 has not proved tractable to X-ray crystallographic analysis, however, structures are available for CDK6 in complex with members of the INK family of CKIs, p16^{INK4a} [48] and p19^{INK4d} [49,48] and in a ternary complex with p18^{INK4c} and cyclin K.[50] Four INK4 family members have been identified, p15^{INK4b}, p16^{INK4a}, p18^{INK4c} and p19^{INK4d} that specifically inhibit CDK4 and CDK6 (reviewed in ref. 51). All the CDK6 complexes are snapshots of the enzyme in various inactive conformations, which leaves a description of the active structure of CDK6 an extrapolation from our current understanding of CDK regulation determined from studies on CDK2.

Structures of CDK6/INK Complexes

The overall fold of CDK6 in both the CDK6/p16^{INK4a} and CDK6/p19^{INK4d} structures is very similar to that of CDK2 (Figs. 6.1 and 6.5). Like CDK2, CDK6 has a bi-lobal architecture with two short, but significant insertions into the fold. CDK6 residues 70-72 add a further turn to the N-terminal domain helix (the C-helix), and residues 86-90 extend β4 and the loop connecting it to β5. In its INK4-bound state, CDK6 is distorted away from the active CDK2 conformation by a relative 15^0 rotation of the N and C-terminal domains around an axis perpendicular to the plane of the catalytic cleft (reviewed in ref. 52). This rotation disrupts the constellation of conserved residues that mediate correct ATP binding. For example it prevents the C-helix from adopting a position equivalent to that seen in the structure of active CDK2. Glu61 (the residue equivalent to CDK2 Glu51) is rotated out of the active site cleft directed towards solvent. Glu99 and Val101 that lie in the hinge region and that, by equivalence to CDK2 residues Glu81 and Leu83 are predicted to form hydrogen bonds with the ATP adenine N6 and N1 atoms respectively, are also displaced.

The CDK6 activation loop can also diverge from that of CDK2. In one structure, the activation loop adopts a fold similar to that observed in monomeric CDK2,[49] but the quality of the electron density map suggests that the sequence is conformationally flexible. The other CDK6/p19^{INK4d} structure (determined from crystals grown under different crystallization conditions) has a highly ordered activation loop adopting a 10 residue β-hairpin.[48] This structure

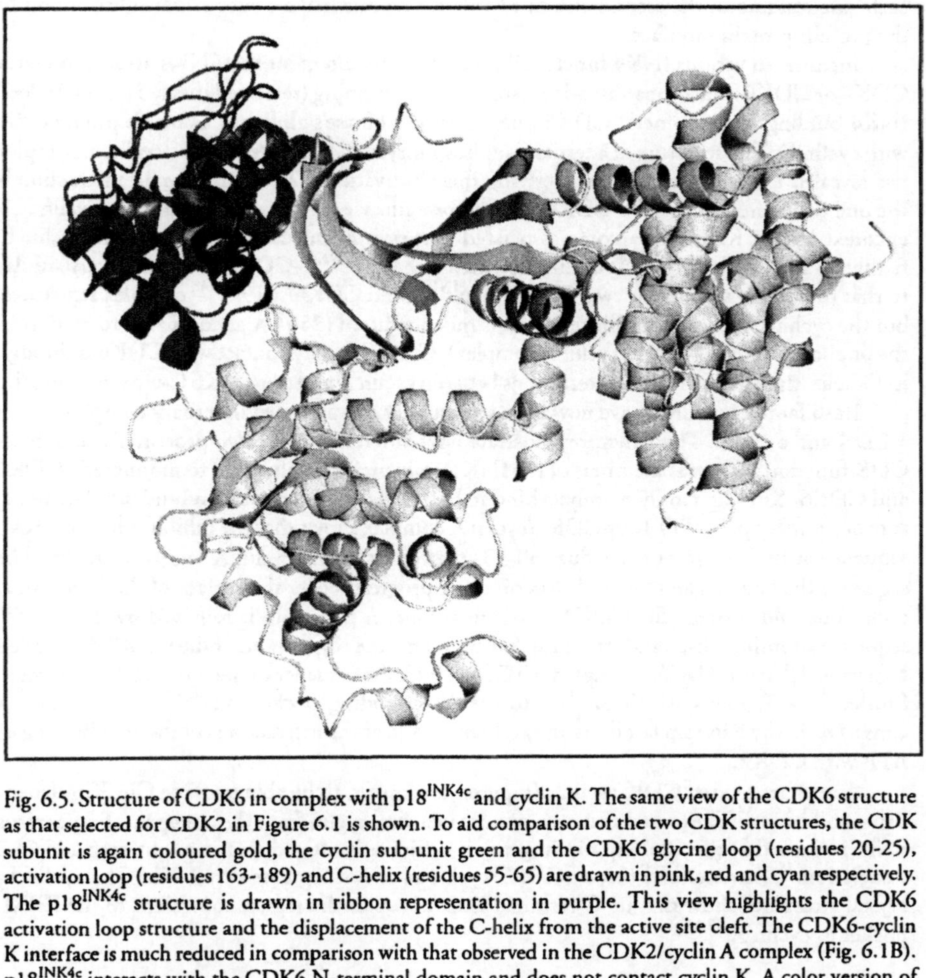

Fig. 6.5. Structure of CDK6 in complex with p18^{INK4c} and cyclin K. The same view of the CDK6 structure as that selected for CDK2 in Figure 6.1 is shown. To aid comparison of the two CDK structures, the CDK subunit is again coloured gold, the cyclin sub-unit green and the CDK6 glycine loop (residues 20-25), activation loop (residues 163-189) and C-helix (residues 55-65) are drawn in pink, red and cyan respectively. The p18^{INK4c} structure is drawn in ribbon representation in purple. This view highlights the CDK6 activation loop structure and the displacement of the C-helix from the active site cleft. The CDK6-cyclin K interface is much reduced in comparison with that observed in the CDK2/cyclin A complex (Fig. 6.1B). p18^{INK4c} interacts with the CDK6 N-terminal domain and does not contact cyclin K. A color version of this figure is included in the insert that precedes Chapter 1.

is stabilised both by hydrogen bonds and hydrophobic interactions with the CDK6 C-terminal lobe and hydrogen bonding contacts to p19^{INK4d}.

The INK4 proteins have been the subject of a number of NMR and X-ray crystallographic studies (reviewed in ref. 53). They contain multiple ankyrin repeats (Fig. 6.5), a structural motif previously identified as a protein-protein interaction domain (reviewed in ref. 54). Each repeat is an L-shaped association of a helical hairpin and two β-strands. There are four ankyrin repeats in p16^{INK4a} (and presumably also in p15^{INK4b} which shows 75% sequence identity to p16^{INK4a}) and five in both p18^{INK4c} and p19^{INK4d}. In the INK4 family the first helical region of the second ankyrin repeat consists of only one helical turn which results in a slight deviation from the consensus ankyrin repeat structure, splaying β–loop 1 away from β-loop 2. p16^{INK4a} and p19^{INK4d} contact both lobes of CDK6, binding to one side of the catalytic cleft opposite to and not overlapping with the cyclin D binding site (Fig. 6.5). Although there are minor differences between the CDK6/p19^{INK4d} and CDK6/p16^{INK4a} structures, in each case residues in the second and third ankyrin repeats make a substantial contribution to the protein-protein interface.

Intrinsic to various INK4 functional models is a requirement that INK4 association with CDK4 or CDK6 is incompatible with stable cyclin D binding (reviewed in ref. 51), and indeed INK4 binding to monomeric CDK6 does affect the kinase's ability to associate productively with cyclin D. The structure of a ternary unphosphorylated CDK6/p18^{INK4c}/cyclin K complex has revealed the structural basis underlying this observation. The CDK6-cyclin interaction is the one that suffers as the two CDK6 binding proteins vie for the enzyme surface. Cyclin K is encoded by the Karposi's sarcoma-associated herpesvirus and is a member of the cyclin D family. In the CDK6/p18^{INK4c}/cyclin K structure the p18^{INK4c}-CDK6 interface is very similar to that previously seen in the two CDK6/ p16^{INK4a} and CDK6/ p19^{INK4d} complex structures, but the cyclin-CDK surface of interaction is much reduced (2504 Å as compared to 3540 Å in the unphosphorylated CDK2/cyclin A complex). Cyclin K's only contact with CDK6 is through its C-helix; there are no further interactions between cyclin K and the CDK6 C-terminal domain.

Both families of CKIs have now been structurally characterised in ternary complexes with a CDK and a cyclin. The structures illustrate how differently these two protein classes inhibit CDK function. Whereas members of the INK family preferentially bind to monomeric CDK4 and CDK6, Kip/Cip family members bind to both the cyclin and CDK subunits.[35] The structure of a ternary p27^{Kip1}/T160pCDK2/cyclin A complex shows that the inhibitor has an ZRXL sequence at its N-terminus (residues 30-33) that binds to the cyclin A recruitment site. The sequence then adopts an α-helical structure as it progresses over the surface of the N-terminal cyclin box fold towards the CDK2 N-terminal lobe. A β-hairpin is followed by the p27^{Kip1} sequence adopting a β-strand structure that supplants the edge strand of the CDK2 5-stranded N-terminal β-sheet. Having disrupted the CDK2 fold, the sequence of the short p27^{Kip1} fragment finishes as a 3_{10}-helix which binds into the ATP binding pocket. p27^{Kip1} Tyr88, which is conserved in the Kip/Cip family, is an excellent mimic of the interactions of the adenine ring of ATP with CDK2.

A number of viral CDK-cyclin complexes are known to be able to evade Cip/Kip-dependent inhibition.[55] The crystal structure of one such viral cyclin from the oncogenic herpes virus saimiri shows that the Cip/Kip binding site is sufficiently altered that it might preclude CKI association.[56] As described below, CDK7/cyclin H is not regulated by CKIs. The cyclin H crystal structures show that its C-terminal helix occupies the region equivalent to the ZRXL pocket in cyclin A.[57,58]

Structural Models for Activation of CDK7

In higher eukaryotes CDK7 is a component of the CDK-activating kinase, (CAK) that phosphorylates CDKs within the activation loop, and of transcription factor IIH (TFIIH), a multi-protein complex that has roles both in nucleotide excision repair of DNA and in RNA polymerase II transcription (reviewed in ref. 59). Single-particle electron microscopy has been employed to reveal the sub-unit organisation within this complex.[60] The *S. pombe* CDK7/cyclin H homologues Mcs6/Mcs2 also function both as a CAK and as components of an RNA polymerase II C-terminal domain (CTD) kinase. In *S. cerevisiae* cells these two functions are carried out by distinct proteins. The *S. cerevisiae* homologue of CDK7, Kin28p is a component of TFIIH. A distinct protein Cak1p, that shares only 20-25% sequence identity to CDK7 and that is active as a monomer phosphorylates both Cdc28p and Kin28p. A monomeric CAK with selectivity for CDK2/cyclin E as a substrate has been identified in human cells.[61]

The cognate cyclin of CDK7 is cyclin H and phosphorylation of CDK7 within its activation segment can stabilise the CDK7/cyclin H pair. Unusually, CDK7 can be phosphorylated at two sites within its activation loop. In the absence of phosphorylation, stable association of CDK7 with cyclin H is promoted by a third protein, MAT1 (ménage a trois). A similar mechanism may exist to promote the formation of other cognate CDK/cyclin pairs. For example, the CKI p21^{Cip1} has been proposed to assist in the formation of CDK4/cyclin D complexes. Measured CDK/cyclin association constants suggest that CDKs have a low intrinsic ability to discriminate between cognate and noncognate cyclin partners. Assembly factors may play a role in promoting the formation of appropriate pairings.

The Structure of Cyclin H

Two crystal structures of human cyclin H have been independently solved to 2.6Å resolution.[57,58] Cyclin H is a shorter cyclin (323 amino acids) than cyclin A and its N-terminus aligns to the start of the truncated cyclin A3 molecule. In the electron density maps for monomeric cyclin H, the C-terminal 36 residues could not be observed[58] suggesting that they are flexible in the absence of a molecular partner. As anticipated, cyclin H shares a duplication of the cyclin box fold and the relative orientation of the two domains is similar to that observed in cyclin A (rmsd of 1.8Å over 132 residues), a result consistent with the notion that cyclin H may bind to and activate its cognate CDK7 in a way similar to that used by cyclin A to activate CDK2. The remarkable difference between cyclins A and H is in the relative positions and contacts made by the N- and C-terminal helices that flank the duplicated cyclin box fold. Both these helices associate with the cyclin H N-terminal cyclin box fold and models suggest that although not at the core of the subunit interface, they may have a role in modulating specificity and activation. Deletion of either helix results in a cyclin capable of binding, but not of activating CDK7.[58] In metazoans Cyclin H can be phosphorylated both in vitro and in vivo by CDK8/cyclin C on Ser5 and Ser304-two modificantions that can repress TFIIH activity both as an activator of transcription and as a CTD kinase.[62] However, in the monomeric cyclin H structures determined to date, both residues lie in disordered regions of the structure.

Docking of the CDK2 structure onto cyclin H to mimic the CDK7-cyclin interaction shows that the cyclin H N-terminal helix would sterically clash with the kinase sub-unit. Kim et al[57] have suggested that a movement of this helix may occur upon binding of cyclin H to CDK7. The cyclin H C-terminal helix occupies the region equivalent to the ZRXL binding pocket in cyclin A thus providing a structural explanation for why cyclin H-dependent kinase activity is not inhibited by members of the Cip/Kip family of CKIs (reviewed in ref. 53).

A Structural Model for CDK7

A sequence comparison of CDK7 (human protein is 346 amino acids, molecular mass 39kDa) and CDK2 shows that the two proteins are very similar (43% identity, 61% similarity), with CDK7 having additional short extensions of 8 residues and 36 residues at the N- and C-terminus respectively. CDK7 is predicted to have the core kinase catalytic fold with greatest similarity to the CDK2 structure. The CDK7 sequence within the conserved glycine-loop motif is GEGQFAT and the enzyme is not subject to regulation by active-site inhibitory phosphorylation. The sequence also contains a recognisable PSTAIREmotif (NRTALRE) preceded by a short (3 amino acid) insertion relative to the CDK2 sequence. CDK7 is phosphorylated within the activation segment on two residues, Ser170 and Thr176, the latter residue equivalent to CDK2 Thr160.

Activation Loop Phosphorylation; Its Role in Stabilizing the CDK7/Cyclin H Complex and in Substrate Selection

The CDK7 and CDK2 activation loop sequences agree well, with no relative insertions and only seven amino acid differences between the DFG and APE motifs. A study using the *X. laevis* CDK7 and cyclin H proteins showed that low affinity CDK7/cyclin H binding was promoted by phosphorylation within the CDK7 activation loop at Ser 170 (Val154 in CDK2), and high affinity binding required phosphorylation ofThr176 (the residue equivalent to CDK2

Thr160).[63] *X. laevis* CDK7/cyclin H singly phosphorylated on Ser170 exhibited almost 90% of the activity of the doubly phosphorylated complex. This result is in contrast to that obtained by Fisher and coworkers using mammalian homologues who showed that phosphorylation of Ser164 in mammalian CDK7 was not sufficient to promote cyclin binding or confer catalytic activity.[64]

A comparison of the T160pCDK2/cyclin A and cyclin H structures suggests that phosphorylation of Ser170 promoting the stability of the CDK7/cyclin H complex might not be unreasonable. In the T160pCDK2/cyclin A structure Val154 (the residue homologous to CDK7 Ser170) is buried at the CDK2/cyclin A interface. A structural alignment of cyclins A and H[57] shows that Arg165 (equivalent to cyclin A Gln313) is one of the residues in the first helix of the second cyclin H cyclin box fold that would be predicted to contact the CDK7 activation loop. Such an additional charge interaction might be sufficient to promote stable cyclin association and the adoption of an activation loop structure that is compatible with substrate recognition and catalysis. The potential structural homology between CDK7 Thr176 and CDK2 Thr160 is readily apparent. The two residues are equivalently positioned within the activation loop sequence and the three positively charged residues (Arg50, Arg126 and Arg150) that coordinate the CDK2 Thr160 phosphate group are conserved in CDK7 (Arg67, Arg142 and Lys166 respectively).

However, CDK2 and CDK7 have very different substrate selectivities. As described, we have structural explanations for the specificity of CDK2/ cyclin A for a proline residue at the P+1 position and for a basic residue preferred at P+3.[15] CDK7 phosphorylates the threonine residue within the consensus sequence YTHEVV, with a slight variation provided by the CDK4 activation loop sequence where a proline is at the P+1 position. Both CDK2 and CDK7 appear to be able to switch specificities. CDK7/cyclin H can act as a proline-directed kinase when acting as a CTD kinase, but as a CAK need not. Similarly, CDK2/cyclin A shows distinct proline-directed specificity for a combinatorial peptide library, but successfully phosphorylates Thr170 of CDK7 in a nonproline directed manner. Clearly these results suggest that there are additional structural determinants of substrate specificity distinct from the primary sequence surrounding the site of phosphorylation. That this may be the case has been elegantly demonstrated by Fisher and colleagues using a chimaeric CDK, where they grafted the activation segment of CDK7 into CDK2.[65] CDK7/cyclin H was now able to phosphorylate this segment but CDK2/cyclin A could not, suggesting that secondary and/or tertiary structural elements are being recognised and can direct specificity. Substrate recruitment could also be important. It is conceivable that recruitment by cyclin A to a CDK7/cyclin H substrate bypasses the proline dependence of CDK2 allowing it to phosphorylate Thr170. Similarly, recruitment of substrates to CDK7 by MAT1 may allow it to acquire specificity for proline-directed phosphorylation sites such as those in the C-terminal domain of RNA polymerase II.

Various members of the CKI families have been reported to inhibit CDK activation by CAK. p21^{Cip1} can inhibit CAK activation of CDK1/cyclin B[66] and p21^{Cip1} and p27^{Kip1}, and p18^{INK4c} have been shown to block CAK-dependent activation of CDK2/cyclin A and CDK6/ cyclin D1 respectively.[67] The T160pCDK2/cyclin A/p27 complex structure shows that p27 has an extended fold that binds to both the cyclin and CDK sub-units and that occupies the CDK catalytic cleft. The structure of only residues 25-93 of p27 was determined and it could be envisaged that the remainder of the protein (198 amino acids in total) could interfere with CAK association. The structure of the CDK6/cyclin K/p18^{INK4c} complex suggests that p18^{INK4c} might not interfere directly with CAK recognition of the CDK activation loop sequence. However, the structure does show that with the CKI bound, the flexible CDK molecule is held in a particular conformation and cannot respond to the cyclin subunit. In the same way, the relative orientation of the CDK N- and C-terminal domains induced by INK binding might not correctly dispose elements of the CAK recognition surface for productive CAK binding. A short peptide derived from p21^{Cip1} that encompasses the recruitment motif (ACRRLFGPVDSE, residues 17-28) has also been shown to inhibit the phosphorylation of CDK2/cyclin E by CDK7/cyclin H.[68] This preliminary result raises the intriguing possibility that CDK/cyclin complexes may recruit CAK activity for their activation.

A Structural Model for MAT1

MAT1, the third component of metazoan CAK (human protein 309 amino acids, molecular mass 36 kDa), acts as an assembly factor for the association of CDK7 and cyclin H.[64,69-71] The protein has an N-terminal Ring Finger motif that is important for optimal TFIIH activity but is not essential for its CDK activating properties.[72] A recent NMR study has revealed the solution structure of this domain (residues 1-65). It has the expected βαββ topology and is conspicuously positively charged.[73]

The MAT1 C-terminal region appears to promote assembly of the CDK and cyclin subunits and stimulate the specific activity of the enzyme, even in the absence of phosphorylation of the activation segment.[72] Furthermore, MAT1 appears to modulate the specificity of CAK, possibly by recruitment.[74] The central portion of MAT1 is predicted to contain a coiled-coil motif and has been shown to be responsible for the association of CAK to TFIIH.[72]

Cak1p/Civ1p

A Structural Model for Cak1p

Database searches show that Cak1p is most closely related to the CDK family (25% identity to Cdc28p). The start of the Cak1p sequence aligns to residue 5 of CDK2 suggesting that the first secondary structural element (β1) of the core kinase catalytic fold is either shortened or missing. The Cak1p sequence is 70 amino acids longer than CDK2 and an alignment shows that the extra sequence is accommodated by several insertions between predicted secondary structural elements. The largest predicted insertion of 29 residues occurs between helices α2 and α3 (helices D and E in PKA nomenclature) in a region called the kinase insert region. Differences are observed in this region, that is located on the opposite face of CDK2 from known regulatory and substrate binding sites, even between closely related CDKs. Shorter inserts (5-15 residues) occur prior to and in the N terminal region of the activation segment and within the loops (L13 and L14) preceding and following α5 (helix G).

Perhaps the most notable feature of the Cak1p sequence is the absence of a glycine-rich motif, a sequence that is involved in ATP-binding and is very strongly conserved in the protein kinase family. In an alignment of 300 protein kinase sequences, only one other protein (a yeast sequence encoding a protein involved in vacuolar protein sorting) lacks all 3 glycines in the canonical GXGXXG motif. Remarkably, deletion of the first 31 amino acids of the Cak1p sequence (which in addition to removing the glycine loop would also remove the putative conserved lysine residue (Lys31, equivalent to CDK2 Lys33) still yielded a functional enzyme.[75]

Cak1p is also conspicuously different in that part of its sequence predicted to encode the C-helix (the PSTAIRE motif in CDK2), presumably reflecting Cak1p's independence from cyclin binding as a requirement for activity. However, the replacement of the PSTAIRE arginine residue (R50 in CDK2) with a phenylalanine must have repercussions for the Cak1p catalytic mechanism. In CDK2 Arg50 (together with Arg126 and Arg150) contributes to the phospho-Thr160 binding pocket. Its absence in Cak1p may be a symptom of the fact that the enzyme is not phosphorylated. Sequence alignments predict that only an arginine equivalent to Arg126 of CDK2 is conserved in Cak1p, Arg150 being replaced by a tyrosine. Taken together the data suggest that Cak1p may adopt an appropriate activation segment conformation for substrate binding and phospho-transfer through hydrophobic packing instead of the charge-mediated interactions seen in CDK2. If so, then phosphorylase kinase may provide a better model for

Cak1p in this part of its structure. In the phosphorylase kinase structure[32] the arginine residue preceding the catalytic aspartate (Asp149) readily compensates the single charge on the glutamate side chain (Glu182) that is the structural equivalent of CDK2 phospho-Thr160. The activation segment of Cak1p contains a glutamate and this residue may similarly stabilise the loop by interacting with the single arginine residue that is conserved in Cak1p that lies next to the catalytic base in the RD motif.

Cak1p has a 7-residue insertion in the activation segment with respect to CDK2 and these extra amino acids must reflect differences in substrate specificity as well as regulatory differences. The activation segment sequences of CDK7 and Cak1p are rather different, even though at least in vitro they can phosphorylate the same or similar substrates. This result may indicate that the two enzymes use different modes of binding. In vitro, Cak1p phosphorylates monomeric CDK2 more effectively than the binary CDK2/cyclin A complex.[14,20] Structures of protein kinase/substrate complexes reveal that the peptide invariably lies in an extended conformation and that the activation segment folds to form pockets to accommodate the side chains responsible for determining specificity.[15,31,32,76] For Cak1p to phosphorylate CDK2 the region around T160 would be expected to adopt such an extended conformation prior to phosphotransfer, a condition that would be disfavoured in the cyclin-bound state.

Some receptor tyrosine kinases, e.g., fibroblast growth factor receptor kinase (FGFRK), platelet derived growth factor receptor kinase (PDGFRK) and insulin receptor kinase (IRK), have significant inserts in the kinase-insert region[2] which have been proposed to be important for receptor dimerisation.[77] In FGFRK this insert contains tyrosine phosphorylation sites. The structure of the inactive, unphosphorylated form of the FGFRK[78] shows that this region is a loop that follows an elongated course over the end of the F helix and the H helix, before running back toward the active cleft to join the E helix. Intriguingly, the Cak1p kinase-insert contains two adjacent tyrosines and it is tempting to speculate that this insert may prove to be a site of protein interaction and/ or a putative site of phosphorylation.

Conclusions

A key element of the regulation of CDK activity is the inherent flexibility of the CDK fold. The CDK2 and CDK6 structures determined to date show how the fold can respond to the binding of multiple regulatory proteins resulting in CDK activation. We now have a structural understanding of CDK activation by cyclin binding and phosphorylation, and of CDK inhibition by binding members of the CKI families. The determination of the structures of complexes containing other CDK/cyclin proteins will be illuminating. CDKs are the hub of many multiprotein complexes and the structural elucidation of these large multi-molecular assemblies will be a challenge for the future.

Abbreviations

CDK2	Monomeric CDK2
T160pCDK2	Monomeric CDK2 phosphorylated on Thr160
T160CDK2/cyclin A	Unphosphorylated binary CDK2/cyclin A complex
T160pCDK2/cyclin A	Binary CDK2/cyclin A complex phosphorylated on CDK2 Thr160
CAK	CDK activating kinase
KAP	Kinase-associated phosphatase
AMP-PNP	Adenylyl imidodiphosphate

Acknowledgments

We would like to thank Martin Noble for his many useful insights and helpful comments during the writing of this Chapter, and for his assistance in the preparation of the figures. We are grateful to our colleagues in the LMB and to the MRC, BBSRC, Oxford University, The Royal Society and The Wellcome Trust for their support.

References

1. De Bondt HL, Rosenblatt J, Jancarik J, Jones HD, Morgan DO et al. Crystal structure of cyclin-dependent kinase 2. Nature 1993; 363:595-602.
2. Hanks SK, Hunter T. The eukaryotic protein kinase superfamily: Kinase (catalytic) domain structure and classification. FASEB J 1995; 9:576-596.
3. Knighton DR, Zheng J, Ten Eyck LF, Ashford VA, Xuong N-H et al. Crystal structure of the catalytic subunit of cyclic adenosine monophosphate-dependent protein kinase. Science 1991; 253:407-413.
4. Jeffrey PD, Russo AA, Polyak K, Gibbs E, Hurwitz J et al. Mechanism of CDK activation revealed by the structure of a cyclinA-CDK2 complex. Nature 1995; 376:313-320.
5. Brown NR, Noble MEM, Endicott JA, Garman EF, Wakatsuki S et al. The crystal structure of cyclin A. Structure 1995; 3:1235-1247.
6. Gibson TJ, Thompson JD, Blocker A, Kouzarides T. Evidence for a protein domain superfamily shared by the cyclins, TFIIB and RB/107. Nucl Acids Res 1994; 22:946-952.
7. Noble MEM, Endicott JA, Brown NR, Johnson LN. The cyclin box fold: protein recognition in cell-cycle and transcription control. Trends Biochem Sci 1997; 22:482-487.
8. Endicott JA, Noble MEM. Structural principles in cell-cycle control: Beyond the CDKs. Structure 1998; 6:535-541.
9. Card GL, Knowles P, Laman H, Jones N, McDonald NQ. Crystal structure of a γ-herpesvirus cyclin-cdk complex. EMBO J 2000; 19:2877-2888.
10. Thuret J-Y, Valay J-G, Faye G, Mann C. Civ1 (CAK In Vivo) a novel CDK-activating kinase. Cell 1996; 86:565-576.
11. Kaldis P, Sutton A, Solomon MJ. The CDK-activating kinase (CAK) from budding yeast. Cell 1996; 86:553-564.
12. Espinoza FH, Farrell A, Erdjument-Bromage H, Tempst P, Morgan DO. A cyclin-dependent kinase-activating kinase (CAK) in budding yeast unrelated to vertebrate CAK. Science 1996; 273:1714-1717.
13. Russo AA, Jeffrey PD, Pavletich NP. Structural basis of cyclin-dependent kinase activation by phosphorylation. Nat Struct Biol 1996; 3:696-700.
14. Brown NR, Noble MEM, Lawrie AM, Morris MC, Tunnah P et al. Effects of phosphorylation of threonine 160 on cyclin-dependent kinase 2 structure and activity. J Biol Chem 1999; 274:8746-8756.
15. Brown NR, Noble MEM, Endicott JA, Johnson LN. The structural basis for specificity of substrate and recruitment peptides for cyclin-dependent kinases. Nat Struct Biol 1999; 1:438-443.
16. Desai D, Wessling HC, Fisher RP, Morgan DO. Effects of phosphorylation by CAK on cyclin binding by CDC2 and CDK2. Mol Cell Biol 1995; 15:345-350.
17. Heitz A, Morris MC, Fesquet D, Cavadore J-C, Doree M et al. Interactions of cyclins with cyclin-dependent kinases: A common interactive mechanism. Biochem 1997; 36:4995-5003.
18. Johnson LN, Noble NEM, Owen DJ. Active and inactive protein kinases: structural basis for regulation. Cell 1996; 85:149-158.
19. Fisher RP, Morgan DO. A novel cyclin associates with MO15/CDK7 to form the CDK-activating kinase. Cell 1994; 78:713-724.
20. Kaldis P, Russo AA, Chou HS, Pavletich NP, Solomon MJ. Human and yeast CDK-activating kinases (CAKs) display distinct substrate specificities. Mol Biol Cell 1998; 9:2545-2560.
21. Solomon MJ, Kaldis P. Regulation of CDKs by phosphorylation. In: Pagano M, ed. Results and Problems in Cell Differentiation: Cell Cycle Control. Heidelberg: Springer-Verlag, 1998:79-109.

22. Kastan MB, Lim D-S. The many substrates and functions of ATM. Nat Rev Mol Cell Biol 2000; 1:179-186.
23. Dasika GK, Lin SC, Zhao S, Sung P, Tomkinson A et al. DNA damage-induced cell cycle checkpoints and DNA strand break repair in development and tumorigenesis. Oncogene 1999; 18:7883-7899.
24. Kastan MB. Checking two steps. Nature 2001; 410:766-767.
25. Pines J. Checkpoint on the nuclear frontier. Nature 1999; 397:104-105.
26. Fauman EB, Cogswell JP, Lovejoy B, Rocque WJ, Holmes W et al. Crystal structure of the catalytic domain of the human cell cycle control phosphatase Cdc25A. Cell 1998; 93:617-625.
27. Reynolds RA, Yem AW, Wolfe CL, Deibel MR, Chidester CG et al. Crystal structure of the catalytic subunit of Cdc25B required for G2/M phase transition of the cell cycle. J Mol Biol 1999; 293:559-568.
28. Barford D, Das AK, Egloff M-P. The structure and mechanism of protein phosphatases: Insights into catalysis and regulation. Ann Rev Biophys Biomol Struct 1998; 27:133-164.
29. Song H, Hanlon N, Brown NR, Noble MEM, Johnson LN et al. Phosphoprotein-protein interactions revealed by the crystal structure of kinase-associated phosphatase in complex with phosphoCDK2. Mol Cell 2001; 7:615-626.
30. Songyang Z, Blechner S, Hoagland N, Hoekstra MF, Piwnica-Worms H et al. Use of an orientated peptide library to determine the optimal substrates of protein kinases. Curr Biol 1994; 4:973-982.
31. Hubbard SR. Crystal structure of the activated insulin receptor tyrosine kinase in complex with peptide substrate and ATP analog. EMBO J 1997; 16:5572-5581.
32. Lowe ED, Noble MEM, Skamnaki VT, Oikonomakos NG, Owen DJ et al. The crystal structure of a phosphorylase kinase peptide substrate complex: kinase substrate recognition. EMBO J 1997; 16:6646-6658.
33. Schulman BA, Lindstrom DL, Harlow E. Substrate recruitment to cyclin-dependent kinase 2 by a multipurpose docking site on cyclin A. Proc Natl Acad Sci USA 1998; 95:10453-10458.
34. Adams PD, Li X, Sellers WR, Baker KB, Leng X et al. Retinoblastoma protein contains a C-terminal motif that targets it for phosphorylation by cyclin-cdk complexes. Mol Cell Biol 1999; 19:1068-1080.
35. Russo AA, Jeffrey PD, Patten AK, Massague J, Pavletich NP. Crystal structure of the $p27^{Kip1}$ cyclin-dependent-kinase inhibitor bound to the cyclin A-Cdk2 complex. Nature 1996; 382:325-331.
36. Lee J-O, Russo AA, Pavletich NP. Structure of the retinoblastoma tumour-suppressor pocket domain bound to a peptide from HPV E7. Nature 1998; 391:859-865.
37. Mittnacht S. Control of pRB phosphorylation. Curr Opin Genet Dev 1998; 8:21-27.
38. Lundberg AS, Weinberg RA. Functional inactivation of the retinoblastoma protein requires sequential modification by at least two distinct cyclin-cdk complexes. Mol Cell Biol 1988; 18:753-761.
39. Harbour JW, Luo RX, Santi AD, Postigo AA, Dean DC. Cdk phosphorylation triggers sequential intramolecular interactions that progressively block Rb function as cells move through G1. Cell 1999; 98:859-869.
40. Hayles J, Aves S, Nurse P. *suc1* is an essential gene involved in both the cell cycle and growth in fission yeast. EMBO J 1986; 5:3373-3379.
41. Pines J. Reaching for a role for the CKS proteins. Curr Biol 1996; 6:1399-1402.
42. Bourne Y, Watson MH, Hickey MJ, Holmes W, Rocque W et al. Crystal structure and mutational analysis of the human CDK2 kinase complex with cell cycle-regulatory protein CksHs1. Cell 1996; 84:863-874.
43. Parge HE, Arvai AS, Murtari DJ, Reed SI, Tainer JA. Human CksHs2 atomic structure: A role for its hexameric assembly in cell cycle control. Science 1993; 262:387-395.
44. Endicott JA, Noble MEM, Garman EF, Brown NR, Rasmussen B et al. The crystal structure of $p13^{suc1}$, a $p34^{cdc2}$-interacting cell cycle control protein. EMBO J 1995; 14:1004-1014.
45. Patra D, Wang SX, Kumagai A, Dunphy WG. The *Xenopus* Suc1/Cks protein promotes the phosphorylation of G_2/M regulators. J Biol Chem 1999; 274:36839-36842.

46. Patra D, Dunphy WG. Xe-p9, a *Xenopus* Suc1/Cks protein, is essential for the Cdc2-dependent phosphorylation of the anaphase-promoting complex at mitosis. Genes Dev 1998; 12:2549-2559.
47. Shteinberg M, Hershko A. Role of Suc1 in the activation of the cyclosome by protein kinase Cdk1/Cyclin B. Biochem Biophys Res Commun 1999; 257:12-18.
48. Russo AA, Tong L, Lee J-O, Jeffrey PD, Pavletich NP. Structural basis for inhibition of the cyclin-dependent kinase cdk6 by the tumour suppressor $p16^{INK4a}$. Nature 1998; 395:237-243.
49. Brotherton DH, Dhanaraj V, Wick S, Brizuela L, Domaille PJ et al. Crystal structure of the complex of the cyclin D-dependent kinase Cdk6 bound to the cell cycle inhibitor $p19^{INK4d}$. Nature 1998; 395:244-250.
50. Jeffrey PD, Tong L, Pavletich NP. Structural basis of inhibition of CDK-cyclin complexes by INK4 inhibitors. Gene Dev 2000; 14:3115-3125.
51. Sherr CJ, Roberts JM. CDK inhibitors: Positive and negative regulators of G_1-phase progression. Gene Dev 1999; 13:1501-1512.
52. Pavletich NP. Mechanisms of cyclin-dependent kinase regulation: structures of Cdks, their cyclin activators, and Cip and INK4 inhibitors. J Mol Biol 1999; 287:821-828.
53. Endicott JA, Noble MEM, Tucker JA. Cyclin-dependent kinases: Inhibition and substrate recognition. Curr Opin Struct Biol 1999; 9:738-744.
54. Bork P. Hundreds of ankyrin-like repeats in functionally diverse proteins: Mobile modules that cross phyla horizontally? Proteins 1993; 17:363-374.
55. Swanton C, Mann DJ, Fleckenstein B, Neipel F, Peters G et al. Herpes viral cyclin/Cdk6 complexes evade inhibition by CDK inhibitor proteins. Nature 1997; 390:184-187.
56. Schulze-Gahmen U, Jung JU, Kim S-H. Crystal structure of a viral cyclin, a positive regulator of cyclin-dependent kinase 6. Structure 1999; 7:245-254.
57. Kim KK, Chamberlin HM, Morgan DO, Kim S-H. Three-dimensional structure of human cyclin H, a positive regulator of the CDK-activating kinase. Nat Struct Biol 1996; 3:849-855.
58. Andersen G, Busso D, Poterszman A, Hwang JR, Wurtz J-M et al. The structure of cyclin H: Common mode of kinase activation and specific features. EMBO J 1997; 16:958-967.
59. Kaldis P. The CDK-activating kinase (CAK): From yeast to mammals. Cell Mol Life Sci 1999; 55:284-296.
60. Schultz P, Fribourg S, Poterszman A, Mallouh V, Moras D et al. Molecular structure of human TFIIH. Cell 2000; 102:599-607.
61. Nagahara H, Ezhevsky SA, Vocero-Akbani AM, Kaldis P, Solomon MJ et al. Transforming growth factor b targeted inactivation of cyclin E: Cyclin-dependent kinase 2 (Cdk2) complexes by inhibition of Cdk2 activating kinase activity. Proc Natl Acad Sci USA 1999; 96:14961-14966.
62. Akoulitchev S, Chuikov S, Reinberg D. TFIIH is negatively regulated by cdk8-containing mediator complexes. Nature 2000; 407:102-106.
63. Martinez A-M, Afshar M, Martin F, J.-C. C, Labbe J-C, Doree M. Dual phosphorylation of the T-loop in CDK7: Its role in controlling cyclin H binding and CAK activity. EMBO J 1997;16:343-354.
64. Fisher RP, Jin P, Chamberlin HM, Morgan DO. Alternative mechanisms of CAK assembly require an assembly factor or an activating kinase. Cell 1995; 83:47-57.
65. Garrett S, Barton WA, Knights R, Jin P, Morgan DO et al. Reciprocal activation by cyclin-dependent kinases 2 and 7 is directed by substrate specificity determinants outside the T loop. Mol Cell Biol 2001; 21:88-99.
66. Smits VAJ, Klompmaker R, Vallenius T, Rijksen G, Mäkelä TP et al. p21 inhibits Thr^{161} phosphorylation of Cdc2 to enforce the G_2 DNA damage checkpoint. J Biol Chem 2000; 275:30638-30643.
67. Aprelikova O, Xiong Y, Liu ET. Both p16 and p21 familiies of cyclin-dependent kinase (CDK) inhibitors block the phosphorylation of cyclin-dependent kinases by the CDK-activating kinase. J Biol Chem 1995; 270:18195-18197.
68. Rank KB, Evans DB, Sharma SK. The N-terminal domains of cyclin-dependent kinase inhibitory proteins block the phosphorylation of cdk2/Cyclin E by the CDK-activating kinase. Biochem Biophys Res Commun 2000; 271:469-473.

69. Devault A, Martinez A-M, Fesquet D, Labbé J-C, Morin N et al. MAT1 (menage a trois) a new RING finger protein subunit stabilising cyclin H-CDK7 complexes in starfish and Xenopus CAK. EMBO J 1995; 14:5027-5036.
70. Tassan J, Schultz SJ, Bartek J, Nigg EA. Cell cycle analysis of the activity, subcellular localisation, and subunit composition of human CAK. J Biol Chem 1994; 127:467-478.
71. Yee A, Nichols MA, Wu L, Hall FL, Kobayashi R et al. Molecular cloning of CDK7-associated human MAT1, a cyclin-dependent kinase activating kinase (CAK) assembly factor. Cancer Res 1995; 55:6058-6062.
72. Busso D, Keriel A, Sandrock B, Poterszman A, Gilead O et al. Distinct regions of MAT1 regulate cdk7 kinase and TFIIH transcription activities. J Biol Chem 2000; 275:22815-22823.
73. Gervais V, Busso D, Wasielewski E, Poterszman A, Egly J-M et al. Solution structure of the N-terminal domain of the human TFIIH MAT1 subunit: New insights into the RING finger family. J Biol Chem 2001; 276:7457-7464.
74. Yankulov KY, Bentley DL. Regulation of CDK7 substrate specificity by MAT1 and TFIIH. EMBO J 1997; 16:1638-1646.
75. Enke DA, Kaldis P, Holmes JK, Solomon MJ. The CDK-activating kinase (Cak1p) from budding yeast has an unusual ATP-binding pocket. J Biol Chem 1999; 274:1949-1956.
76. Knighton DR, Zheng J, Ten Eyck LF, Xuong N-H, Taylor SS et al. Structure of a peptide inhibitor bound to the catalytic subunit of cyclic adenosine monophosphate-dependent protein kinase. Science 1991; 253:414-420.
77. Weiss A, Schlessinger J. Switching signals on or off by receptor dimerisation. Cell 1998; 94:277-280.
78. Mohammadi M, Schlessinger J, Hubbard SR. Structure of the FGF receptor tyrosine kinase domain reveals a novel autoinhibitory mechanism. Cell 1996; 86:577-587.

Chapter 7

Activation of CDKs by CAK:

CAK in TFIIH

Anne Keriel and Jean-Marc Egly

Introduction

TFIIH was originally identified as a basal transcription factor involved in protein-coding genes transcription.[1,2] The systematic cloning of its nine subunits revealed however that TFIIH also participates in two other fundamental cell processes: cell cycle regulation and repair of damaged DNA. Because TFIIH is involved in more than one cellular process, mutations in some of its subunits can lead to human syndromes : Xeroderma pigmentosum, Trichothiodystrophy and Cockayne syndrome.[3,4] These genetic disorders are caracterized by complex phenotypes with large ranges of pleiotropic symptoms, including UV-light sensitivity, cancer susceptibilities and developmental abnormalities.

In this study, we will describe the function(s) of a ternary subcomplex of TFIIH, named "CAK" (CDK-activating kinase) because of its ability to activate CDKs by phosphorylation, keeping in mind that TFIIH possesses additional enzymatic activities involved in both RNA synthesis and DNA repair.

CAK as Part of TFIIH

The mammalian TFIIH is a multiprotein complex of nine subunits, ranging from 89 kDa to 32 kDa (Fig. 7.1), that can be resolved in two main functional subcomplexes: the coreTFIIH, composed of five subunits (XPB, p62, p52, p44 and p34), and the CAK (CDK-activating kinase), composed of CDK7, cyclin H and MAT1[5,6] (for details see Chapter 1). The remaining XPD subunit can be found either associated with the core or with the kinase complex[7,8] and is believed to anchor the CAK to the coreTFIIH.[9] TFIIH possesses three intrinsic enzymatic activities: the cyclin-dependent kinase CDK7 and two ATPase/helicases of opposite polarities attributed to the XPB and XPD subunits. The two helicases are responsible for catalysing the DNA opening that takes place either around the initiation site, during transcription, or around a lesion, to allow DNA repair[10-12] (cf. paragraph 2). The kinase activity of CDK7 is directed towards the cyclin-dependent kinases (CDKs), in agreement with the initial definition of the CAK complex, but also towards RNA polymerase II (RNA pol II), general transcription factors as well as transcription activators and repressors (cf. paragraph 3). Interestingly, the preference of CAK for one of these two classes of substrates depends on whether it resides in TFIIH or exists as a free complex, suggesting that free CAK is devoted to cell cycle regulation whereas, when associated to the coreTFIIH, it participates to TFIIH function(s) in the transcription reaction.[13,14]

The CDK-Activating Kinase (CAK), edited by Philipp Kaldis.

A

	TFIIH Subunit	Molecular Weight (KDa)	Enzymatic Activity/Function	*S. cerevisiae*	
Core	XPB	89	ATPase/3'->5' helicase, DNA opening	Rad25/Ssl2	
	p62	62	nd	Tfb1*	
	p52	52	XPB anchoring	Tfb2*	
	p44	44	Stimulation of XPD helicase activity	Ssl1*	
	p34	34	nd	Tfb4	
	XPD	80	ATPase/5'->3' helicase, CAK anchoring	Rad3*	
CAK	cdk7	40	cyclin-dependant kinase, CTD kinase	Kin28	TFIIK
	cyclin H	35	regulation of cdk7	Ccl1	TFIIK
	MAT1	32	regulation of cdk7	Tfb3/Rig2*	

B

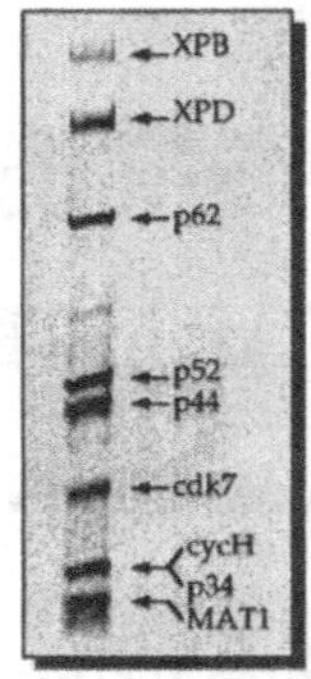

Fig. 7.1. TFIIH composition and subunits functions. (A) Characteristics of the nine subunits of the DNA repair/transcription factor TFIIH. (B) Silver-stained SDS-PAGE gel of highly purified human TFIIH factor. nd : not determined; asterisks point out to the subunits composing the yeast coreTFIIH subcomplex.

Cloning of the genes encoding the nine subunits of TFIIH revealed a highly conserved structure and function from yeast to human[15] (Fig. 7.1). However, although yeast TFIIH has the same composition as the mammalian counterpart, some differences exist:

1. the yeast coreTFIIH subcomplex contains Tfb1p (p62), Tfb2p (p52), and Ssl1p (p44), but also Rad3p (XPD) and Tfb3p (MAT1) which are absent in the human core factor,[16]
2. in contrast to their human homologues, the yeast Ssl2p (XPB) helicase and Tfb4p (p34) are not present in the core subcomplex and[17]
3. the kinase subcomplex (called TFIIK) contains only Kin28p (CDK7) and Ccl1p (cyclin H)[18,19] and does not exhibit a CDK-activating kinase (CAK) activity.[20]

In *S. cerevisiae*, the CDK-activating function has been devoted to Cak1p/Civ1p (for details see Chapter 2), a kinase that is not associated to TFIIH.[21-24]

The differences observed between human and yeast TFIIH also appeared from their quaternary organization. The three-dimensional (3D) model designed from electron microscopy analysis, shows that the human TFIIH is organized into a ring-like structure, with a hole whose size is suitable to accommodate a double-stranded DNA molecule, and from which an almost spherical bulge of protein density protrudes out[25] (Fig. 7.2A). The human recombinant coreTFIIH subcomplex forms a circular architecture that can be superimposed on the ring found in human TFIIH, suggesting that the CAK constitutes the bulge appended to the ring-like structure.[25] The quaternary organization of TFIIH subunits can be partially inferred from immunolabeling combined with coimmunoprecipitation experiments (Fig. 7.2A). The CDK7 kinase was shown to be located in the protruding domain. The p44 subunit is located within the ring structure, at the basis of the protruding protein density, and is flanked on either side by the XPB and XPD helicases. It is interesting to note that the location of XPD, close to the CAK-containing bulge, is consistent with the proposal that it could bridge (or stabilize) the interaction between the CAK and the coreTFIIH. The resolution of the 2D crystallographic structure of the yeast coreTFIIH subcomplex, which contains the helicase Rad3p (XPD), Ssl1p (p44), Tfb1p (p62), Tfb2p (p52), and Tfb3p (MAT1), was also reported.[26] In this complex, Tfb1p, Tfb2p, and Tfb3p form a ring-like structure to which is appended Rad3p, via a bridge created by Ssl1p (Fig. 7.2B, in purple). A plausible fit of the crystal structure of yeast coreTFIIH

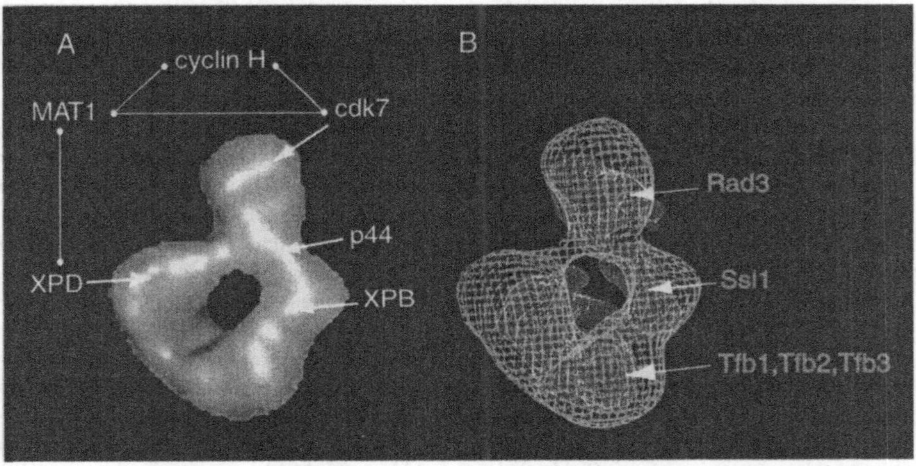

Figure 7.2. Quaternary organizations of human TFIIH and yeast coreTFIIH. A) Surface representation of the 3D model reconstruction of human TFIIH from electron microscopy analysis. The positions of subunits CDK7, XPD, XPB, and p44, as inferred from immunolabeling experiments, are indicated by the arrows. The cyclin H, MAT1, and p34 subunits are tentatively positioned according to well established pairwise subunit-subunit interaction studies. (Adapted from Chang et al, Cell 2000; 102:609-613.) B) Quaternary organizations of human TFIIH and yeast coreTFIIH. Superimposition of the 2D crystal structure of yeast coreTFIIH with human TFIIH (grey). Assignments of density regions to subunits in the yeast structure, either Rad3p, Ssl1p, or Tfb1p/Tfb2p/Tfb3p, are indicated. (Adapted from Chang et al, Cell 2000; 102:609-613.) A color version of this figure is included in the insert that precedes Chapter 1.

to the human TFIIH particle reconstruction could be proposed (Fig 7.2B), pointing to differencies and similarities in size, shape, and arrangement of subunits. The human and yeast studies are not fully compatible because:

1. the yeast coreTFIIH subcomplex presents a large nonring density area,
2. the rings observed in each strcuture are of different size and subunits composition and
3. the localization of the Rad3p/XPD and the Tfb3p/MAT1 subunits in the two structures are in clear disagreement (Fig. 7.2B).

The discrepancies in the subunit composition and quaternary organization between the yeast and human TFIIH subcomplexes probably reflect some subtle differences concerning the role of their various subunits in the DNA repair and transcription activities of TFIIH.

The Role of CAK in DNA Repair

DNA repair pathways are fundamental cellular processes for protecting cells against genetic damage produced by various genotoxic agents. They proceed essentially either through the direct replacement of the nucleotide modified by a lesion or through the excision of an oligonucleotide carring the damage and the subsequent replacement by a newly synthesised intact DNA (reviewed in 27). Among the various DNA repair pathways, nucleotide excision repair (NER) plays an essential role in cell survival by removing the major UV-induced DNA lesions. Defects in this DNA repair pathway generate three human genetic disorders: Xeroderma pigmentosum (XP), Trichothiodystrophy and Cockayne syndrome. XP is a highly cancer-prone skin disorder which symtoms are caused by exposure to sun-light and can be accounted for the deficiency in NER. Eight complementation groups of XPs (XP-A to XP-G + XP-V) have been identified; two of them result from mutations in the XPB and XPD helicases of TFIIH.

Recently, the identification of the various components of the NER pathway, the mutation of which generate the DNA repair disorders, allowed the reconstitution of the NER reaction in vitro.[28] NER proceeds through different steps : damage recognition, opening of the DNA around the lesion, incision on each side, excision of the ~30-mer oligonucleotide bearing the damage, replacement of the excised fragment by resynthesis and ligation (Fig. 7.3).[29] Briefly, XPC/HR23B and XPA bind the damaged DNA, followed by the opening of the DNA around the lesion, promoted by the two helicases subunits of TFIIH, XPD and XPB. The RPA protein, with the help of XPA, would then maintain the DNA opened, allowing cleavage by the two XPG and XPF-ERCC1 nucleases. Once the damaged fragment is excised, de novo DNA synthesis is achieved by the DNA polymerase δ/ε under the control of PCNA (Proliferating Cell Nuclear Antigen) and RCF (replication factor C) and the reaction is terminated by DNA ligase I.[28,30,31]

The involvement of TFIIH in NER was emphasized by immunodepletion experiments, with anti-p34 antibodies or anti-p52 antibodies, which led to the inhibition of NER activity both in vitro and in vivo.[32,33] These observations raised the question of the role of each TFIIH subunit in NER. According to the UV-sensitivity phenotype of the different mutant cells in yeast and human, at least five TFIIH subunits could be considered as repair proteins: the XPB/Rad25p and XPD/Rad3p helicases, as well as p62/Tfb1p,[34,35] p44/Ssl1p[35,36] and p52/Tfb2p.[15] To the opposite, yeast strains carrying mutations either in Kin28p or in Tfb3p exhibit only a mild, if any, UV-sensibility and no decrease in DNA damage repair.[19,37] Altogether, these results minimize the role of CAK in NER. In agreement with this, a CAK-depleted TFIIH was shown to be still active in a reconstituted yeast[38] or mammalian[39] in vitro system. However, at least two studies imply the CAK component of TFIIH in NER. First, microinjection of antibodies raised against CDK7 results in a drop of NER in vivo.[5] Second, a tfb3 temperature-sensitive mutant yeast strain was shown to exhibit both a severe UV-sensitivity and a defect in NER in vitro.[40] Moreover, using a reconstituted excision assay, it was demonstrated that the presence of CAK inhibits the reaction.[41] This inhibition was observed only in the presence of an ATP-regenerating system. Addition of a specific kinase inhibitor overcame this inhibition, suggesting that it is the kinase activity of CAK, and not its physical presence, that is involved in this negative regulation.

In the light of the above results, it remains difficult to design a scenario in which TFIIH would work either in transcription or in NER as a function of its structural state (absence or presence of CAK, modification of some of its subunits,...). Nevertheless, two distinct TFIIH complexes have been found in yeast: one of them contains the kinase and the cyclin subunits and is involved in transcription; the other one, which forms the "repairosome complex" with other NER components and seems therefore devoted to NER, lacks TFIIK.[40,42] The transition from one form to another would enable competition to occur between transcription and NER.[42] This model is, however, not applicable to mammalian cells: first, eventhough mammalian TFIIH easily separates into several subcomplexes upon purification, the existence of the coreTFIIH subcomplex has never been demonstrated in cells; second, microinjection of anti-CDK7 antibodies inhibits NER in vivo,[5] suggesting that the 9-subunits TFIIH normally participates in NER. It is thus believed that, in mammals, the two functions of TFIIH do not involve two distinct TFIIH complexes ; the switch between these two functions would rather be the result of intramolecular events induced by DNA damage. A model was proposed in which CDK7 (or CAK) negatively regulates the NER reaction by phosphorylating some NER components,[41] and is itself inhibited in response to DNA damage. Indeed, the kinase activity of the TFIIH purified from UV-irradiated mammalian cells is dramatically inhibited.[6] Moreover, treatment of HeLa cell extracts with serine-threonine phosphatase inhibitors prevents incision of the damaged oligonucleotide during the NER reaction.[43] In this model, XPG, which was found to interact physically with and be phosphorylated by TFIIH[44,45] is a good target for such inhibition.

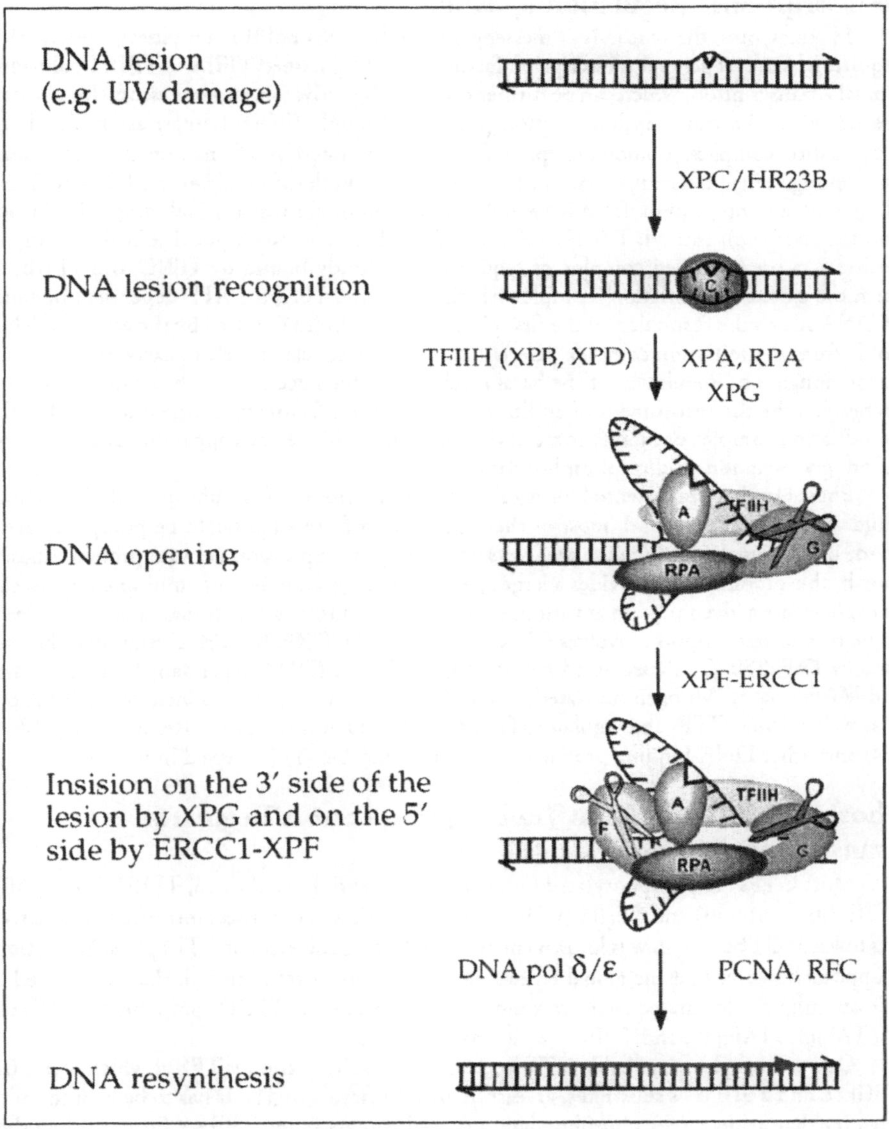

Fig. 7.3. Schematic representation of the nucleotide excision repair (NER) pathway. The involvement of the various NER proteins (XPC/HR23B, TFIIH, RPA, XPA, XPG, XPF-ERCC1, DNA polymerase, PCNA, RCF and DNA ligase I) are shown at each step of the reaction (damage recognition, DNA opening, incision, excision, resynthesis and ligation), as described in the text. A color version of this figure is included in the insert that precedes Chapter 1.

As a conclusion, it has to be noticed that, eventhough CDK7 might not be directly involved in the NER reaction, it likely plays a crucial role in the cell cycle arrest induced upon genotoxic attacks, allowing DNA repair. When DNA is damaged, TFIIH would activate p53 by phosphorylation, which would further induce the CDK-inhibitor $p21^{WAF1/CIP1}$ gene expression, leading to cell cycle arrest.[46]

Transcriptional Regulation by CAK

In eukaryotes, the synthesis of messenger RNA by RNA pol II is a multistep process that requires at least six general transcription factors (or GTFs, named TFIIA to TFIIH), allowing « basal » transcription, which can be further regulated by activators and repressors (for a review see ref. 47). The transcription reaction proceeds through different steps: assembly of the preinitiation complex, promoter opening, first phosphodiesterbond formation, promoter clearance, elongation and termination. The first step consists in the assembly of the different GTFs, along with a nonphosphorylated RNA pol II, at the promoter (usually including a TATA-box and the start site). Indeed, TFIIH, TFIIA, TFIIF, TFIIE and RNA pol II join the promoter (either as a preassembled complex or sequentially), already bound by TFIID and TFIIB, to form the closed "preinitiation" complex which is inactive. Then the ATP-dependent opening of DNA allows the formation of the first phosphodiester bond followed by the escape of RNA pol II from the initiation complex, referred as to "promoter clearance", to carry on with transcript elongation. Regulation of the basal machinery, which occurs in a chromatin context, is orchestered by the recruitment of additional transcription factors that can either stabilize the preinitiation complex or modify some of the components of the basal apparatus through acetylation, glycosylation, and/or phosphorylation.

Phosphorylation is essential for regulation of transcription; first, phosphorylation is both rapid and reversible; second, most of the transcription factors appear to be phosphorylated; third, many signal transduction pathways affecting gene expression activate protein kinases; fourth, this modification provides a large spectrum of regulation because multiple kinases can phosphorylate a given protein at various sites, within its different functional domains. Indeed, regulation of transcription involves CDK7, CDK8, and CDK9, but also several other kinases such as $TAF_{II}250$ (TBP-associated factor 250), DNA-PK (DNA-dependant Protein Kinase) and MAP kinases (Mitogen Activated Protein Kinases). Their substrates include the RNA pol II as well as some GTFs and regulatory factors such as nuclear receptors (RARα, RARγ, ERα, AR) and other DNA-binding proteins (GAL4, p53 and Oct-1) (reviewed in ref. 48).

Phosphorylation of Basal Transcription Factors Regulates Transcription

Most GTFs are phosphorylated (at least in vitro) and three of them, TFIID ($TAF_{II}250$), TFIIF (its α subunit) and TFIIH (CDK7), have been shown to possess intrinsic kinase activities towards GTFs. Very few is known concerning the consequences of GTFs phosphorylation. It appears however that the global repression of RNA pol II transcription that is observed at mitosis, might be facilitated by the reversible phosphorylation of TFIID components ($TAF_{II}20$/15, $TAF_{II}31$, $TAF_{II}80$, and TBP)[49] (reviewed in ref. 50).

One of the most remarkable GTF in terms of phosphorylation is TFIIH which uses TBP, TFIIEα, and TFIIFα as substrates, at least in vitro (reviewed in 51). It has to be kept in mind however that evidence for phosphorylation of TBP (in vivo) and TFIIFα (in vitro) by other kinases than TFIIH, have been obtained. Moreover, eventhough TFIIF phosphorylation was found to reduce the initiation and the elongation kinetic,[52,53] a direct link between phosphorylation of TFIIFα by TFIIH and the regulation of TFIIF activity has not been provided yet.

TFIIH is also a substrate for other cyclin-dependent kinases. The p62 and CDK7 subunits of TFIIH are phosphorylated in vivo by CDK1/cyclin B, resulting in a silencing of transcription and an inhibition of CDK7 kinase activity during mitosis.[54,55] In addition, phosphorylation of cyclin H by CDK8/cyclin C represses both the ability of TFIIH to support activated transcription and its CTD kinase activity causing, as a global consequence, inhibition of cellular proliferation.[56]

Regulation of Transcription and RNA pol II CTD Phosphorylation

The carboxy-terminal domain (CTD) of the largest subunit of RNA pol II, which contains up to 52 repeats of the heptapeptide Tyr1-Ser2-Pro3-Thr4-Ser5-Pro6-Ser7, represents an ideal substrate for serine/threonine kinases. Many questions remain unanswered concerning the role of this remarkable domain. It is nevertheless clear that its phosphorylation/dephosphorylation follows some steps of the transcription reaction: initiation and elongation, respectively. It has been indeed demonstrated that RNA pol II engaged in the elongation step of transcription is hyperphosphorylated, whereas initiation requires a hypophosphorylated RNA pol II (reviewed in ref. 57). Deletions of the CTD domain in vivo are lethal[58] and its presence is required for in vitro transcription. It has to be pointed out however that CTD requirement for in vitro transcription depends on the nature of the promoter: a CTD-truncated RNA pol II is able to catalyse RNA synthesis from the Adenovirus major late promoter (AdMLP) but not from the TATA-less dihydrofolate reductase (DHFR) promoter.[59]

The kinase of TFIIH uses CTD as a substrate.[60] A null mutation in CDK7, either in vivo[61] or in vitro,[12] leads to a dramatic inhibition of CTD phosphorylation. In addition, both CTD phosphorylation and RNA pol II transcription are reduced at the restriction temperature in a yeast *kin28* (the counterpart of the human CDK7) temperature-sentitive mutant.[20,37] Not only the kinase itself, but also cyclin H and MAT1, the two partners of CDK7 within CAK, are involved in the phosphorylation process. Indeed, a conditional *ccl1* yeast mutant (the counterpart of human cyclin H) shows a strong decrease of the in vivo level of CTD phosphorylation.[62] Mutation in MAT1 also affects both CTD phosphorylation and basal transcription. It was suggested that disruption of the RING-finger motif of MAT1 does not further allow optimal positioning of some of the components of the basal transcription machinery on the promoter, causing a drop in the affinity of CDK7 for its CTD substrate and/or a decrease of RNA synthesis.[63]

Although not really proven, it seems that CTD phosphorylation would favor the transition from initiation to elongation by allowing the RNA pol II to be released from TFIID, which remains bound to its TATA-box target site.[60,64] In addition, it has to be noticed that transcription from the DHFR promoter needs CAK kinase activity,[59] whereas both basal and activated transcription from the AdMLP requires neither the enzymatic activity of CAK[59,61] nor its presence within TFIIH.[7,8,13] In fact, in vitro CTD phosphorylation requires neither promoter opening nor formation of the first phosphodiester bond[12] (Fig. 7.4). It remains that CAK stimulates significantly the formation of the first phosphodiester bond,[65] but this involves only a structural contribution of the complex and not its enzymatic activity.[66]

Although not absolutely required for initiation of transcription from every promoter in vitro, phosphorylation of the CTD likely serves as a control panel for transcription and post-transcriptional events in vivo. Besides controlling the progression of RNA pol II through the transcription cycle, CTD phosphorylation and dephosphorylation has also been directly linked to other gene expression related processes such as premRNA processing (capping, splicing, poly(A)site cleavage) [reviewed in 67], as well as ubiquitin-dependent degradation of RNA pol II.[68]

CAK is Involved in Activated Transcription

Since transcription regulators interact directly, or via cofactors, with the basal transcription machinery, it was hypothesized that some kinases present within this machinery, such as CDK7, TAFII250 or TFIIFα, would participate in the regulation of transcription. CDK7 (or its yeast homologue Kin28p) was indeed found to phosphorylate the activation domains of both E2F1[69] and the yeast transcription factor GAL4,[70] the DNA binding domains of GAL4[70] and Oct1[71] and the activation and C-terminal multi-functional domains of p53.[72,73] As a

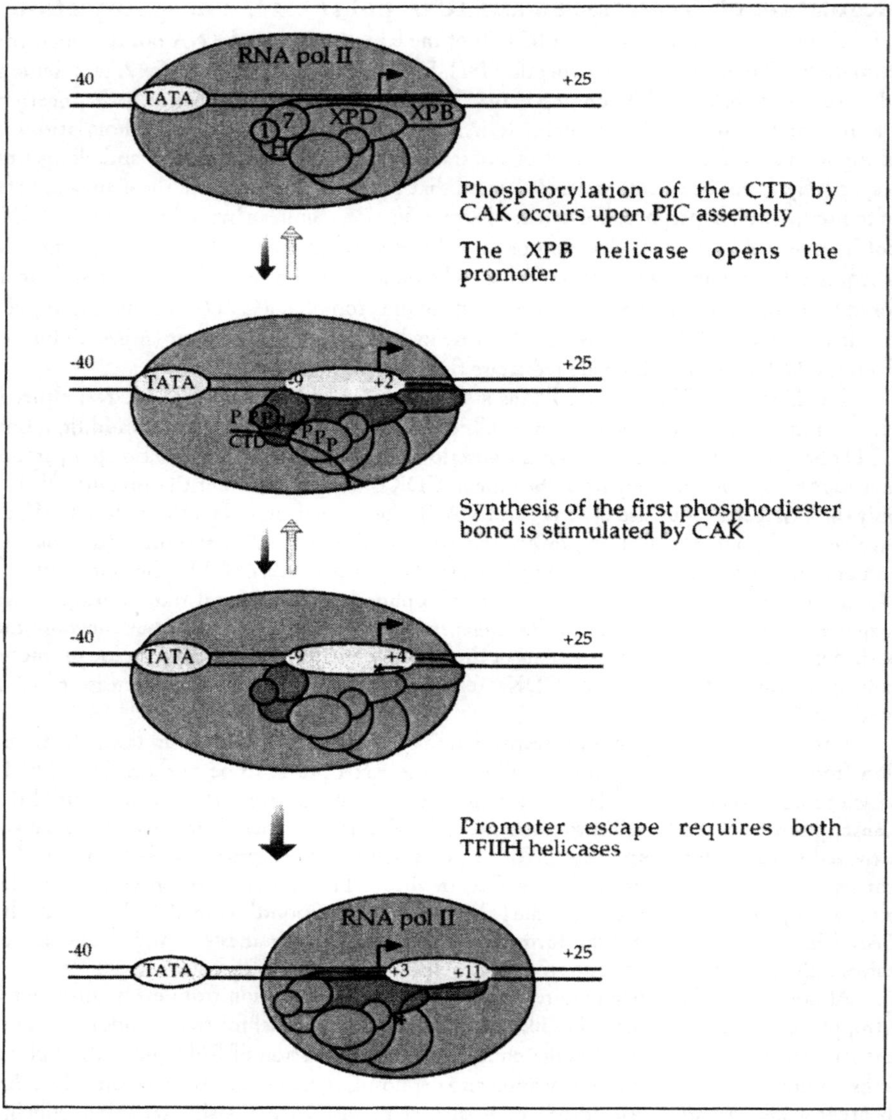

Fig. 7.4. Proposed role(s) of TFIIH subunits in transcription initiation. The involvement of the CAK subcomplex or of the two helicases subunits of TFIIH into the different steps of the initiation of transcription are shown. A color version of this figure is included in the insert that precedes Chapter 1.

consequence, one could observe variations in gene expression levels. Interestingly, the retinoic acid receptors (RAR) α and γ, and the estrogen receptor alpha (ERα) are phosphorylated by CDK7 in vitro, as well as in vivo.[74-76] Such phosphorylation by TFIIH parallels an increased level of target genes expression. Recently, a physical interaction was also detected between the androgen receptor (AR) and TFIIH.[77] RARs, ERα and AR belong to the superfamily of nuclear receptors which share common conserved domains: a ligand-independent activation domain named AF-1, encompassing the phosphorylation site for CDK7, a DNA-binding domain (DBD), a ligand-binding domain (LBD) and an activation domain that is inducible by cognate ligands, AF-2 (reviewed in 78). The nuclear receptors regulate target gene expression by binding to specific DNA sequences, named "response elements", either as homo- or heterodimers.

The mechanism of action of all these transcriptional regulators is still unclear. However, it seems that the phosphorylation of AF-1 domains of RARs and ERα by CDK7 is essential for transactivating target genes expression.[74-76] It cannot be excluded that phosphorylation might regulate either the binding of the receptor to its responsive element and/or its dimerization. It is not known whether, once phosphorylated, these receptors directly stimulate some components of the transcription machinery (basal factors and/or coactivators). It can be hypothesized that, upon interaction with TFIIH and further phosphorylation, these DNA binding proteins could stimulate some steps of the transcription reaction such as CTD phosphorylation, DNA opening, promoter clearance, or recruitment and/or stimulation of the elongation machinery. Indeed, phosphorylation of the DNA-binding domain of p53 by CDK7, which stimulates in vitro the transcription of some genes, enhances its affinity for its responsive DNA-element.[73] Other studies suggest that such phosphorylation modulate interactions with proteins such as MDM-2, CBP/p300 or TFIID.[79] Nevertheless, to date, it has not been demonstrated that CDK7 phosphorylation could play a role in the transactivation of transcription mediated by p53 in vivo. Finally, an additional mechanism of transcription factor regulation is provided by phosphorylation of the activation domain of E2F-1 by CDK7, which is responsible for targeting E2F-1 to ubiquitin-dependent proteolysis, thus affecting E2F-1 activity at the protein level.[69]

Conclusions and Perspectives

With the discovery of the multiple functions of TFIIH, this factor has emerged as a pivotal component of the cellular life, likely coordinating cell cycle progression with transcription and DNA repair. The activity of CDK7, originally identified as the kinase responsible for CDK activation within a free CAK complex, had thus to be analyzed in the scope of the multisubunit TFIIH complex and in its roles in transcription and DNA repair.

There are still unanswered questions concerning the role of the kinase of TFIIH in basal transcription: does it regulate RNA synthesis through stimulation of some of the first steps of the transcription initiation, or does it simply rhythm two of the main steps of the transcription reaction: initiation and elongation? Does CDK7 coordinate some RNA processing events such as splicing or polyadenylation through phosphorylation of the CTD of RNA pol II? If we extend our observations to transcriptional activation, does CDK7 play a role in regulation of specific genes and, as thus, is under the control of external stimuli and does it function in relation with some chromatin assembly/disassembly processes?

We may then wonder how the kinase of TFIIH might be regulated upon genotoxic attack, knowing that TFIIH is an essential factor of DNA repair. In such a situation, could one emphasized that CDK7 exert a role at this level and, if so, which one? At this stage of our knowledge, the simplest hypothesis considers the CAK subcomplex (and/or its kinase activity) as a key element for the transition of TFIIH roles from transcription to DNA repair. However, further studies are definitely required to challenge this model.

These are some of the questions that are worthwhile to take into consideration if one would try to further understand the mechanisms that underlie regulation and coupling of two fundamental cell processes: transcription and DNA repair.

Abbreviations

CDK(s):	cyclin-dependent kinase(s)
CAK:	CDK-activating kinase
NER:	nucleotide excision repair
RNA pol II:	RNA polymerase II
CTD:	carboxy-terminal domain of the largest subunit of RNA polymerase II
GTF(s):	general transcription factors
TFII:	transcription factor of class II genes

Acknowledgments

We are grateful to all members of our group for fruitful discussions and critical reading of the manuscript. This work was supported by grants from the Ministère de l'Eduction Nationale et de la Recherche Scientifique et Technique, from the Association pour la Recherche sur le Cancer (Grant ARC9083), and from the Human Frontier Program (RG-193/97) and EEC grants (QLG1-1999). A.K. was a recipient of a fellowship from the Ministère de l'Eduction Nationale et de la Recherche Scientifique et Technique.

References

1. Conaway JW, Conaway RC. A multisubunit transcription factor essential for accurate initiation by RNA polymerase II. J Biol Chem 1989; 264:2357-2362.
2. Gerard M, Fischer L, Moncollin V et al. Purification and interaction properties of the human RNA polymerase B(II) general transcription factor Btf2. J Biol Chem 1991; 266:20940-20945.
3. Taylor E, Broughton D, Botta E et al. *Xeroderma pigmentosum* and trichothiodystrophy are associated with different mutations in the XPD (Ercc2) repair/transcription gene. Proc Natl Acad Sci USA. 1997; 94:8658-8663.
4. Botta E, Nardo T, Broughton BC et al. Analysis of mutations in the XPD gene in Italian patients with trichothiodystrophy: Site of mutation correlates with repair deficiency, but gene dosage appears to determine clinical severity. Am J Hum Genet 1998; 63:1036-1048.
5. Roy R, Adamczewski JP, Seroz T et al. The MO15 cell cycle kinase is associated with the TFIIH transcription-DNA repair factor. Cell 1994; 79:1093-1101.
6. Adamczewski JP, Rossignol M, Tassan JP et al. MAT1, CDK7 and cyclin H form a kinase complex which is UV light-sensitive upon association with TFIIH. EMBO J 1996; 15:1877-1884.
7. Drapkin R, Le Roy G, Cho H et al. Human cyclin-dependent kinase-activating kinase exists in three distinct complexes. Proc Natl Acad Sci 1996; 93:6488-6493.
8. Reardon JT, Ge H, Gibbs E et al. Isolation and characterization of two human transcription factor IIH (TFIIH)-related complexes: Ercc2/CAK and TFIIH. Proc Natl Acad Sci USA 1996; 93:6482-6487.
9. Coin F, Marinoni JC, Rodolfo C et al. Mutations in the XPD helicase result in Xp and Ttd phenotypes, preventing the interaction of XPD with the p44 subunit of TFIIH. Nat Genet 1998; 20:184-188.
10. Evans E, Fellows J, Coffer A et al. Open complex formation around lesion during nucleotide excision repair provides a structure for cleavage by human XPG protein. EMBO J 1997; 16:625-638.
11. Mu D, Wakasugi M, Hsu DS et al. Characterization of reaction intermediates of human excision repair nuclease. J Biol Chem 1997; 272:28971-28979.
12. Tirode F, Busso D, Coin F et al. Reconstitution of the transcription factor TFIIH: Assignment of the functions for the three enzymatic subunits, XPB, XPD and CDK7. Mol Cell 1999; 3:87-95.
13. Rossignol M, Kolb-CheynelI, Egly JM. Substrate specificity of the CDK-activating kinase (CAK) is altered upon association with TFIIH. EMBO J 1997; 16:1628-1637.

14. Yankulov K, Bentley DL. Regulation of CDK7 substrate specificity by MAT1 and TFIIH. EMBO J 1997; 16:1638-1646.
15. Feaver WJ, Henry NL, Wang Z et al. Genes for Tfb2, Tfb3, and Tfb4 subunits of yeast transcription/repair factor IIH. Homology to human cyclin-dependent kinase activating kinase and IIH subunits. J Biol Chem 1997; 272:19319-19327.
16. Svejstrup JQ, Feaver WJ, LaPointe L et al. RNA polymerase transcription factor IIH holoenzyme from yeast. J Biol Chem 1994; 269:28044-28048.
17. Feaver WJ, Svejstrup JQ, Bardwell L et al. Dual roles of a multiprotein complex from *S. cerevisiae* in transcription and DNA repair. Cell 1993; 75:1379-1387.
18. Svejstrup JQ, Feaver WJ, Kornberg RD. Subunits of yeast RNA polymerase II transcription factor TFIIH encoded by the Ccl1 gene. J Biol Chem 1996; 271:643-645.
19. Faye G, Simon M, Valay JG et al. Rig2, a ring finger protein that interacts with the Kin28/Ccl1 CTD kinase in yeast. Mol Gen Genet 1997; 255:460-466.
20. Cismowski MJ, Laff GM, Solomon MJ et al. Kin28 encodes a C-terminal domain kinase that controls mRNA transcription in *Saccharomyces cerevisiae* but lack cyclin-dependant kinase-activating activity. Mol Cell Biol 1995; 15:2983-2992.
21. Kaldis P, Sutton A Solomon MJ. The CDK-activating kinase (CAK) from budding yeast. Cell 1996; 86:553-564.
22. Espinoza FH, Farrell A, Erdjument-Bromage H et al. A cyclin-dependent kinase-activating kinase (CAK) in budding yeast unrelated to vertebrate CAK. Science 1996; 273:1714-1717.
23. Thuret JY, Valay JG, Faye G et al. Civ1 (CAK in vivo), a novel CDK-activating kinase. Cell 1996; 86:565-576.
24. Kaldis P. The CDK-activating kinase (CAK): From yeast to mammals. Cell Mol Life Sci 1999; 55:284-96.
25. Schultz P, Fribourg S, Poterszman A et al. Molecular structure of human TFIIH. Cell 2000; 102:599-607.
26. Chang WH, Kornberg RD. Electron crystal structure of the transcription factor and DNA repair Complex, core TFIIH. Cell 2000; 102:609-613.
27. Friedberg EC, Walker GC, Siede G. DNA repair and Mutagenesis. Washington, DC: ASM Press, 1995.
28. Aboussekhra A, Biggerstaff M, Shivji MKK et al. Mammalian DNA nucleotide excision repair reconstituted with purified components. Cell 1995; 80:859-868.
29. Wood RD. Nucleotide excision repair in mammalian cells. J Biol Chem 1997; 272:23465-23468.
30. Mu D, Park CH, Matsunaga T et al. Reconstitution of human DNA repair excision nuclease in a highly defined system. J Biol Chem 1995; 270:2415-2418.
31. Guzder SN, Habraken Y, Sung P et al. Reconstitution of yeast nucleotide excision repair with purified rad proteins, replication protein A, and transcription factor TFIIH. J Biol Chem 1995; 270:12973-12976.
32. Humbert S, van Vuuren H, Lutz Y et al. p44 and p34 subunits of the Btf2/TFIIH transcription factor have homologies with Ssl, a yeast protein involved in DNA repair. EMBO J 1994; 13:2393-2398.
33. Marinoni JC, Roy R, Vermeulen W et al. Cloning and characterization of p52, the fifth subunit of the core of the transcription/DNA repair factor TFIIH. EMBO J 1997; 16:1093-1102.
34. Matsui P, DePaulo J, Buratowski S. An interaction between the Tfb1 and Ssl1 subunits of yeast TFIIH correlates with DNA repair activity. Nucleic Acids Res 1995; 23:767-772.
35. Wang Z, Buratowski S, Svejstrup JQ et al. The yeast Tfb1 and Ssl1 genes, which encode subunits of transcription factor IIH, are required for nucleotide excision repair and polymerase II transcription. Mol Cell Biol 1995; 15:2288-2293.
36. Yoon H, Miller SP, Pabich EK et al. Ssl1, a suppressor of a His4 5'-Utr stem-loop mutation, is essential for translation initiation and affects UV resistance in yeast. Genes Dev 1992; 6:2463-2477.

37. Valay JG, Simon M, Dubois MF et al. The Kin28 gene is required both for RNA polymerase II mediated transcription and phosphorylation of the Rpb1p CTD. J Mol Biol 1995; 249:535-544.
38. Sung P, Guzder SN, Prakash L et al. Reconstitution of TFIIH and requirement of its DNA helicase subunits, Rad3 and Rad25, in the incision step of nucleotide excision repair. J Biol Chem 1996; 271:10821-10826.
39. Mu D, Hsu DS, Sancar A. Reaction mechanism of human DNA repair excision nuclease. J Biol Chem 1996; 271:8285-8294.
40. Feaver WJ, Huang W, Gileadi O et al. Subunit interactions in yeast transcription/repair factor TFIIH. Requirement for Tfb3 subunit in nucleotide excision repair. J Biol Chem 2000; 275:5941-5946.
41. Araujo SJ, Tirode F, Coin F et al. Nucleotide excision repair of DNA with recombinant human proteins: Definition of the minimal set of factors, active forms of TFIIH, and modulation by CAK. Genes Dev 2000; 14:349-359.
42. Svejstrup JQ, Wang Z, Feaver WJ et al. Different forms of TFIIH for transcription and DNA repair: Holo-TFIIH and a nucleotide excision repairosome. Cell 1995; 80:21-28.
43. Ariza RR, Keyse SM, Moggs JG et al. Reversible protein phosphorylation modulates nucleotide excision repair of damaged DNA by human cell extracts. Nucleic Acids Res 1996; 24:433-440.
44. Iyer N, Reagan MS, Wu J et al. Interactions involving the human RNA polymerase II transcription factor/nucleotide excision repair complex TFIIH, the nucleotide excision repair protein XPG, and cockayne syndrome group B (Csb) protein. Biochemistry 1996; 35:2157-2167.
45. Habraken Y, Sung P, Prakash S et al. Transcription factor TFIIH and DNA endonuclease Rad2 constitute yeast nucleotide excision repair factor 3: Implications for nucleotide excision repair and cockayne syndrome. Proc Natl Acad Sci USA 1996; 93:10718-10722.
46. Ko LJ, Prives C. p53: Puzzle and paradigm. Genes Dev 1996; 10:1054-1072.
47. Orphanides G, Lagrange T, Reinberg D. The general transcription factors of RNA polymerase II. Genes Dev 1996; 10:2657-2683.
48. Riedl T, Egly JM. Phosphorylation in transcription: The CTD and more. Gene Expr 2000; 9:3-13.
49. Segil N, Guermah M, Hoffmann A et al. Mitotic regulation of TFIID: Inhibition of activator-dependent transcription and changes in subcellular localization. Genes Dev 1996; 10:2389-2400.
50. Gottesfeld JM, Forbes DJ. Mitotic repression of the transcriptional machinery. Trends Biol Sci 1997; 22:197-202.
51. Frit P, Bergmann E, Egly JM. Transcription factor IIH: A key player in the cellular response to DNA damage. Biochimie 1999; 81:27-38.
52. Kitajima S, Chibazakura T, Yonaha M et al. Regulation of the human general transcription initiation factor TFIIF by phosphorylation. J Biol Chem 1994; 269:29970-29977.
53. Rossignol M, Keriel A, Staub A et al. Kinase activity and phosphorylation of the largest subunit of TFIIF transcription factor. J Biol Chem 1999; 274:22387-22392.
54. Akoulitchev S, Reinberg D. The molecular mechanism of mitotic inhibition of TFIIH is mediated by phosphorylation of CDK7. Genes Dev 1998; 12:3541-3550.
55. Long JJ, Leresche A, Kriwacki RW et al. Repression of TFIIH transcriptional activity and TFIIH-associated CDK7 kinase activity at mitosis. Mol Cell Biol 1998; 18:1467-1476.
56. Akoulitchev S, Chuikov S, Reinberg D. TFIIH is negatively regulated by CDK8-containing mediator complexes. Nature 2000; 407:102-106.
57. Dahmus ME. Reversible phosphorylation of the C-terminal domain of RNA polymerase II. J Biol Chem 1996; 271:19009-19012.
58. Ossipow V, Tassan JP, Nigg EA et al. A mammalian RNA polymerase II holoenzyme containing all the components required for promoter-specific transcription initiation. Cell 1995; 83:137-146.
59. Akoulitchev S, Mäkelä TP, Weinberg RA et al. Requirement for TFIIH kinase activity in transcription by RNA polymerase II. Nature 1995; 377:557-560.

60. Lu H, Zawel L, Fisher L et al. Human general transcription factor IIH phosphorylates the C-terminal domain of RNA polymerase II. Nature 1992; 358:641-645.
61. Mäkelä TP, Parvin JD, Kim J et al. A kinase-deficient transcription factor TFIIH is functional in basal and activated transcription. Proc Natl Acad Sci USA 1995; 92:5174-5178.
62. Valay JG, Dubois MF, Bensaude O et al. Ccl1, a cyclin associated with protein kinase Kin28, controls the phosphorylation of RNA polymerase II largest subunit and mRNA transcription. C R Acad Sci Paris 1996; 319:183-189.
63. Busso D, Kériel A, Sandrock B et al. Distinct regions of MAT1 regulate CDK7 kinase and TFIIH transcription activities. J Biol Chem 2000; 275:22815-22823.
64. Usheva A, Maldonado E, Goldring A et al. Specific interaction between the nonphosphorylated form of RNA polymerase II and the tata-binding protein. Cell 1992; 69:871-881.
65. Seroz T, Perez C, Bergmann E et al. p44/Ssl1, the regulatory subunit of the XPD/Rad3 helicase plays a crucial role in the transcriptional activity of TFIIH. J Biol Chem 2000; 275:33260-33266.
66. Bradsher J, Coin F, Egly JM. Distinct roles for the helicases of TFIIH in transcript initiation and promoter escape. J Biol Chem 2000; 275:2532-2538.
67. McCracken S, Fong N, Yankulov K et al. The C-terminal domain of RNA polymerase II couples mRNA processing to transcription. Nature 1997; 365:357-361.
68. Mitsui A, Sharp PA. Ubiquitination of RNA polymerase II large subunit signaled by phosphorylation of carboxyl-terminal domain. Proc Natl Acad Sci USA 1999; 96:6054-6059.
69. Vandel L, Kouzarides T. Residues phosphorylated by TFIIH are required for E2f-1 degradation during S-phase. EMBO J 1999; 18:4280-4291.
70. Hirst M, Kobor MS, Kuriakose N et al. Gal4 is regulated by the RNA polymerase II holoenzyme-associated cyclin-dependent protein kinase Srb10/Cdk8. Mol Cell 1999; 3:673-678.
71. Inamoto S, Segil N, Pan ZQ et al. The cyclin-dependent kinase-activating kinase (CAK) assembly factor, MAT1, targets and enhances CAK activity on the POU domains of octamer transcription factors. J Biol Chem 1997; 272:29852-29858.
72. Ko LJ, Shieh SY, Chen X et al. p53 is phosphorylated by CDK7-cyclin H in a p36MAT1-dependent manner. Mol Cell Biol 1997; 17:7220-7229.
73. Lu H, Fisher RP, Bailey P et al. The CDK7-cyclin H-p36 complex of transcription factor IIH phosphorylates p53, enhancing its sequence-specific DNA binding activity in vitro. Mol Cell Biol 1997;17:5923-5934
74. Rochette-Egly C, Adam S, Rossignol M et al. Stimulation of RAR alpha activation function Af-1 through binding to the general transcription factor TFIIH and phosphorylation by CDK7. Cell 1997; 90:97-107.
75. Bastien J, Adam-Stitah S, Riedl T et al. TFIIH interacts with the retinoic acid receptor gamma and phosphorylates its Af-1-activating domain through CDK7. J Biol Chem 2000; 275:21896-21904.
76. Chen D, Riedl T, Washbrook E et al. Activation of estrogen receptor alpha by S118 phosphorylation involves a ligand-dependent interaction with TFIIH and participation of CDK7. Mol Cell 2000; 6:127-137.
77. Lee K, Duan HO, Chang C. From androgen receptor to the general transcription factor TFIIH. Identification of Cdk activating kinase (CAK) as an androgen receptor NH(2)-terminal associated coactivator. J Biol Chem 2000; 275:9308-9313.
78. Chambon P. A decade of molecular biology of retinoic acid receptors. FASEB J 1996; 10:940-954.
79. Meek W. Mechanisms of switching on p53: A role for covalent modification? Oncogene 1999; 18:7666-7675.

Chapter 8

The Evolution of CDK-Activating Kinases

Ji Liu and Edward T. Kipreos

Introduction

Cyclin-dependent kinases (CDKs) are essential regulators of the cell cycle and transcription. In the budding yeast *Saccharomyces cerevisiae* and the fission yeast *Saccharomyces pombe*, a single CDK (Cdc28p or its ortholog Cdc2, respectively) catalyzes all major cell cycle transitions.[1,2] In higher eukaryotes, there has been an expansion in the number of CDKs that regulate the cell cycle. For example, in mammals, CDK4, CDK6, and CDK3 regulate the G1–S phase transition; CDK2 controls entry into S phase and DNA replication; and CDK1 (CDC2) is essential for mitosis.[3-7]

CDKs also regulate transcription. In budding yeast, Kin28p, Srb10p, and Ctk1p regulate mRNA synthesis by phosphorylating the carboxyl-terminal domain (CTD) of RNA polymerase II.[8-11] The budding yeast CDK Sgv1p also functions to regulate transcription, but its substrates are unknown.[12] In metazoa, the ortholog of Sgv1p, CDK9, is a component of the positive transcription elongation factor b (pTEFb) that promotes productive RNA Pol II transcription elongation by phosphorylating the CTD.[13,14]

CDK activity is highly regulated in cells. There are four major regulatory mechanisms: 1) most CDKs require binding to cyclin proteins to become active; 2) CDK activity is inhibited by the binding of cyclin-dependent kinase inhibitors (CKIs); 3) phosphorylation of conserved residues in the ATP-binding pocket of the CDK inhibits its activity; and 4) phosphorylation of a conserved residue in the T-loop of CDKs is required for the activation of most CDKs.

The activating phosphorylation of CDKs is catalyzed by a CDK–activating kinase (CAK). CAK phosphorylates a conserved serine (Ser) or threonine (Thr) site in the T-loop of the CDKs. We will refer to this site as Thr 160 (T160) based on its location in human CDK2. When not phosphorylated, the T-loop blocks the entrance of the CDK active site cleft to prevent the binding of protein substrates.[15] Phosphorylation on T160 induces a conformational change in the CDK, resulting in enhanced CDK–cyclin interaction and substrate binding.[16] Both cell cycle CDKs such as CDK1, CDK2, and CDK4, and transcription CDKs such as CDK7 and Kin28p have been shown to require CAK phosphorylation to be active.[17-23] A loss of CAK activity leads to cell cycle arrest as well as transcription defects.[22-24]

The Evolution of CAKs

In this chapter, we address the evolution of CAKs in the context of the evolution of the entire CDK-family, to which known CAKs belong. The focus of the phylogenetic analyses described in this article is the assignment of orthology. Orthologous genes are direct descendents from the same ancestral gene and are only separated from each other by speciation events. Biochemical and cellular functions of orthologs are often conserved across species, and

The CDK-Activating Kinase (CAK), edited by Philipp Kaldis.

therefore orthologs are candidates for functional tests. Further, understanding orthology provides information on how central biological processes have evolved over time, e.g., whether a particular cellular process is ancient and performed by orthologous genes in diverged organisms or whether different genes have evolved to perform the process. In the case of CAK evolution, it is clear that certain CAK pathways have not been conserved across all eukaryotes. The approach described in this article is to identify all CDK-family members in completed genomes, and then assign orthology based on significant associations in phylogenetic trees.

Two Categories of CAKs

Two categories of CAKs have been found in eukaryotes. The first category includes two orthologous genes, budding yeast *CAK1* and its fission yeast ortholog *Csk1*. The second category contains orthologous genes from metazoa, *CDK7*, and fission yeast, *Mcs6*. Members of both categories of CAKs belong to the extended CDK family.[25]

Cak1p and Csk1 are diverged CDK family members.[17-19] When comparing the conserved kinase domain,[26] Cak1p and Csk1 only share 20% sequence identity with Cdc28p, the prototype CDK, while most yeast CDKs are more than 35% identical to Cdc28p. Both Cak1p and Csk1 are monomeric CAKs, that is, they are constitutively active without the need for a cyclin partner.[17-19] Further, they do not require phosphorylation of the T160 site to be active.[17-19] In fact, neither Cak1p nor Csk1 has a Ser or Thr at the T160 position.

CDK7 and Mcs6 are more conventional CDKs. They share 40% sequence identity with CDC28 in the kinase domain, they have Thr or Ser at their T160 sites, and they both need cyclin partners to be active.[27-30] Monomeric unphosphorylated CDK7 cannot efficiently bind to its cyclin partner, cyclin H; it can only bind cyclin H after having been phosphorylated on its T160 site, or alternatively, after binding of the assembly factor MAT1.[31] CDK7/Mcs6 family members function not only as CAKs but also as essential transcription factors. They are components of the general transcription factor TFIIH and function as activating CTD kinases.[32] Kin28p, CDK7's budding yeast ortholog, is required for the transcription of 87% of all budding yeast genes.[33]

Different organisms use CAKs in different ways. In budding yeast, Cak1p is the sole known CAK and is required for the phosphorylation of both the cell cycle CDK Cdc28p and the transcription CDK Kin28p.[17,18,22,23] Kin28p, which is the ortholog of CDK7 and Mcs6, does not possess CAK activity.[8,34,35] In fission yeast, the monomeric CAK Csk1 phosphorylates and activates Mcs6. Csk1 and Mcs6 work redundantly to phosphorylate the cell cycle CDK Cdc2.[19,36] In metazoa, *Drosophila* Cdk7 was shown to function as a CAK for the M phase cell cycle CDK Cdc2, but not for the S phase CDK Cdk2/Cdc2c.[20] No *CAK1/Csk1* ortholog has been found in metazoa.

Yeast CDK and CAK Orthologs

Budding yeast Cak1p and fission yeast Csk1 were initially thought to be unrelated kinases, since they share little sequence homology.[19,36] However, by phylogenetic analysis, we observed that Cak1p and Csk1 group together with significant bootstrap support (Fig. 8.1).[25] Both Cak1p and Csk1 have diverged considerably not only from other CDKs but also from other kinases, with both genes missing a number of key residues conserved in the eukaryotic protein kinase superfamily.[17] These observations suggest that *CAK1* and *Csk1* are rapidly evolving genes. In phylogenetic analyses, fast-evolving genes can group together solely because of their shared dissimilarity relative to slow-evolving taxa. This misleading convergence is called long–branch attraction.[37] To study the true phylogenetic relationship of Cak1p and Csk1, we used four methods to circumvent long branch attraction: First, we included a variety of the closest related outgroup taxa in our analysis, as the inclusion of such taxa helps to minimize long

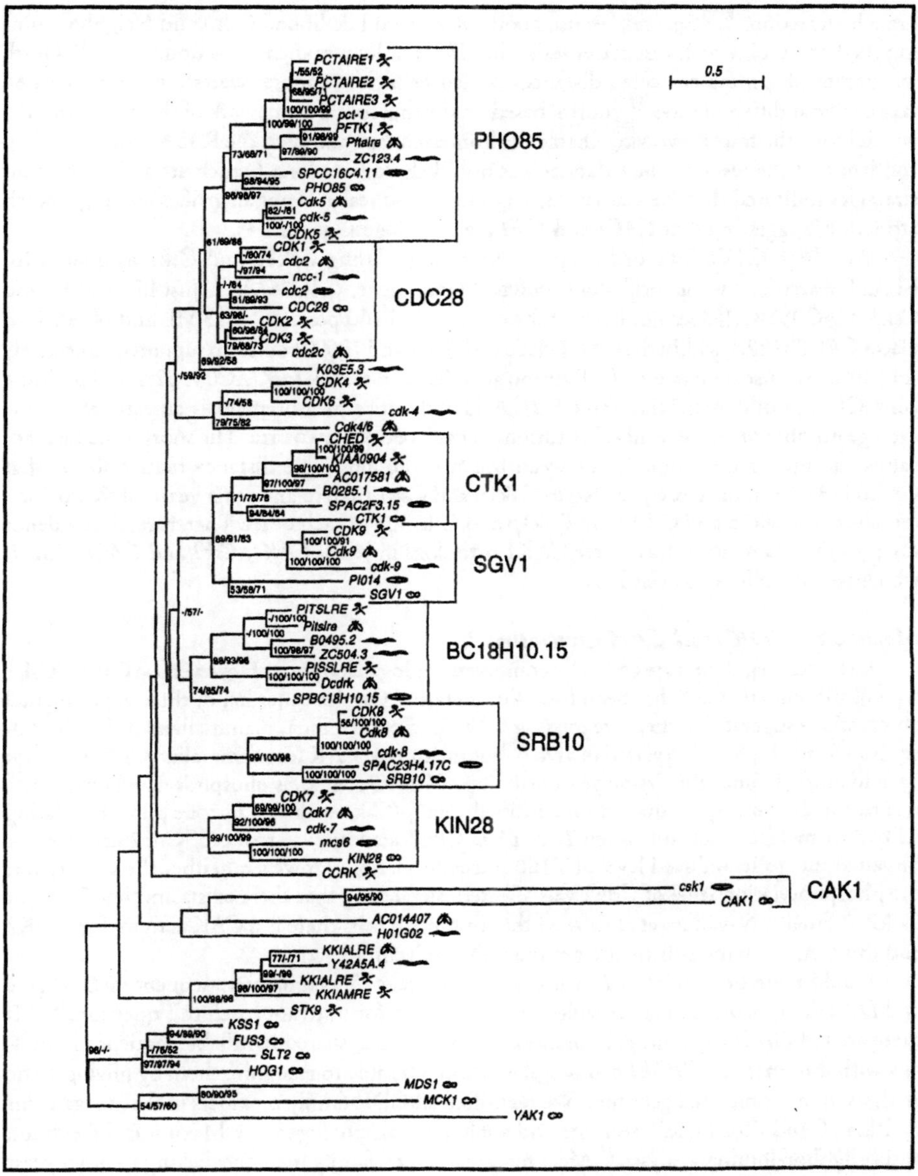

Fig. 8.1. Gamma-rate corrected neighbor-joining phylogeny of *S. cerevisiae, S. pombe, D. melanogaster, C. elegans,* and human extended CDK family members.[25] The amino acid data set was modeled onto one invariable and eight gamma-rates with the program PUZZLE to produce the pair-wise distances that were used to create the NJ (neighbor-joining analysis) tree. Branch lengths are proportional to the estimated number of amino acid substitutions; scale bar indicates amino acid substitutions per site. Bootstrap support values above 50% are given at branch nodes and are derived from ML (Maximum Likelihood analysis, left), NJ of uncorrected data set (center), and gamma-rate corrected NJ (right) analyses, separated by slash marks. Species are denoted by cartoon. Ancestral clades are denoted by brackets on the right. Reprinted with permission from Liu J and Kipreos E, Mol Biol Evol 2000; 17:1061-1074. © Society for Molecular Biology and Evolution.

branch attraction.[38,39] Second, we used both Maximum Likelihood (ML) and Neighbor-Joining (NJ) analyses, which can better resist the effects of long branch attraction.[37,39,40,41] Third, we employed gamma-corrected distances, which facilitates a more accurate phylogeny when taxa evolve at different rates.[42] Fourth, based on the theoretical framework of the S-F method,[43] we excluded the fastest-evolving characters from each taxa and used the RASA program[38,44] to confirm that the resultant new dataset was not likely to have long-branch attraction. All four strategies indicated that the Cak1p-Csk1 grouping was real and not an artifact of long-branch attraction, suggesting that *CAK1* and *Csk1* are orthologous genes.

Are there *CAK1/Csk1* orthologs in other fungi? Using Cak1p and Csk1 as queries for BLAST searches,[45] we identified a *Candida albicans* gene, *CAC05182.1*. Just like Cak1p and Csk1, CAC05182.1 does not have a Ser or Thr at its T160 position. Our ML and NJ analysis places CAC05182.1 within the CAK1 clade with 99% and 100% bootstrap supports, respectively (Fig. 8.2). We also generated a JTT amino acid distance matrix for CAC05182.1 and budding yeast CDK family members (Table 8.2). A JTT distance measurement is an estimate of the average number of amino acid substitutions per site between two taxa. The shorter the distance value, the higher the similarity between the two proteins. The distance matrix shows that CAC05182.1 is more closely related to Cak1p and Csk1 than to any other yeast CDKs. In fact, it is closer to Cak1p and Csk1 than Cak1p and Csk1 are to each other. Therefore, the evidence suggests that there are at least three CAK1 orthologs in yeast: *CAK1, Csk1* and *CAC05182.1*, which we now refer to as *CaCAK1*.

Metazoan CDK and CAK Orthologs

Cell cycle regulators are generally conserved among eukaryotes. However, no Cak1p/Csk1 type of monomeric CAK has been found in metazoa. On the other hand, three experimental observations suggest that there are missing CAKs in metazoa. First, mammalian CDK7's T160-equivalent site is phosphorylated in vivo,[31] but the in vivo CAK for mammalian CDK7 has not been identified. Since the fission yeast ortholog of CDK7, Mcs6, is phosphorylated by Csk1,[19] it is reasonable to suspect that an unidentified Cak1p/Csk1 ortholog may be phosphorylating CDK7 in metazoa. Second, when *Drosophila* Cdk7 activity was reduced, Cdc2 activity was impaired due to its reduced level of T160 phosphorylation. However, neither the activity nor the phosphorylation level of Cdk2 was affected, suggesting that there exists another CAK for Cdk2.[20] Finally, Nagahara et al showed that in human cells there is a CAK activity for CDK2 and this CAK activity is distinct from the CDK7/Cyclin H activity.[24]

Could there be *CAK1/Csk1* orthologs in metazoa? The completely sequenced *C. elegans* and *D. melanogaster* genomes provided an opportunity for us to approach this question.[46,47] If there are *CAK1/Csk1* orthologs in metazoa and if they still share identifiable sequence homology with the yeast *CAK1/Csk1* orthologs, we should be able to recognize them by phylogenetic analysis in the completed genomes. We identified all CDK family members in these organisms by BLAST and PROFILE[48] searches and analyzed their phylogeny by Maximum Likelihood and Neighbor-Joining analyses.[25] Most metazoan CDK family members belong to other yeast CDK clades with significant bootstrap supports, indicating that they are paralogs instead of orthologs of *CAK1/Csk1*. We observed that no metazoan CDK-family members group with the CAK1 clade with significant bootstrap supports (Fig. 8.1).

A true metazoan CAK1 ortholog may be too divergent to group with the yeast *CAK1/Csk1* orthologs with high bootstrap support. There are several orphan metazoan CDK family members that do not group with any of the clades with significant bootstrap scores, making them potential *CAK1/Csk1* orthologs. For example, the *Drosophila* protein AC017707 and the *C. elegans* protein H01G02.2 branched at the base of the CAK1 clade, and human CCRK branched nearby (Fig. 8.1).

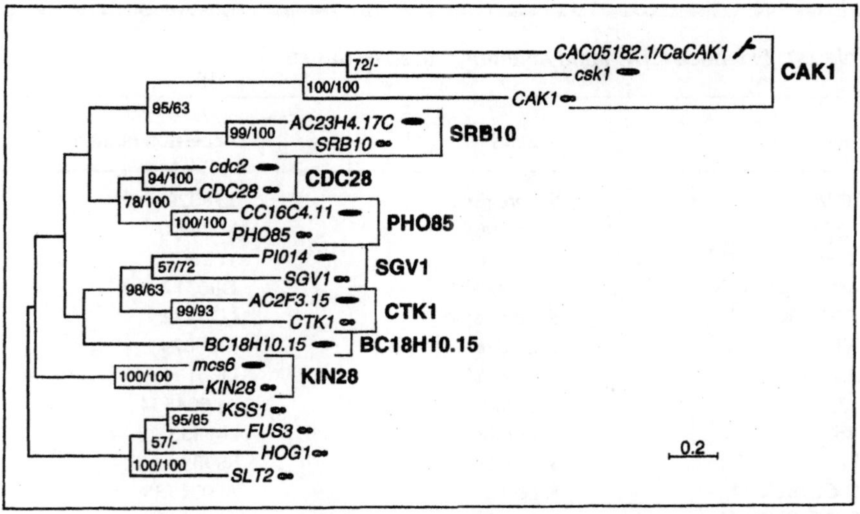

Fig. 8.2. Neighbor-Joining (NJ) and Maximum Likelihood (ML) phylogenies of *S. cerevisiae*, *S. pombe*, and *Candida albicans* CDK family members. Branch lengths are proportional to the estimated number of amino acid substitutions; scale bar indicates amino acid substitutions per site. Bootstrap scores were obtained from 1000 replicates for the NJ analysis and 10,000 replicates for the ML analysis; scores above 50% are given at branch nodes (ML on the left and NJ on the right, separated by slash marks). Species are denoted by cartoon. Ancestral clades are denoted by brackets on the right. For a description of the methods see ref. 25. The sequence alignment used for this analysis can be obtained upon request from the authors.

Sequence analyses of these proteins indicated that H01G02.2 and CCRK share one characteristic with *CAK1/Csk1* orthologs: they do not have Ser or Thr at their T160 sites. H01G02.2 has an aspartic acid (D) at the T160 position, which could mimic the structural effect of a phosphorylated Ser or Thr to produce a constitutively active kinase.[49,50] CCRK has a T-loop deletion and therefore presumably does not have the physical blockage towards substrate binding that the T-loop imposes. Thus, H01G02.2 and CCRK may not need to be phosphorylated by a CAK to become active.

However, a recently identified mouse CCRK ortholog, PNQLARE (GenBank acc. no. AAF89089), has an intact T-loop with a Thr at the T160 position. This finding suggests that the vertebrate ancestor of CCRK had a T-loop and may therefore have required activating phosphorylation.

Research in our lab suggests that *C. elegans* H01G02.2 is not likely to function as a CAK. First, a deletion allele of *H01G02.2* that is a molecular null (obtained from the *C. elegans* Gene Knockout Consortium) is phenotypically wild type and fertile without obvious cell cycle defects. Second, H01G02.2 does not seem to be working redundantly with C. elegans *cdk-7*, as inactivating both *H01G02.2* and *cdk-7* by RNAi did not cause a more severe defect than inactivating *cdk-7* alone. Furthermore, while *Csk1* can complement a *cak1* mutant,[19] ectopic expression of *H01G02.2* failed to complement budding yeast *cak1* mutants.[25]

Arabidopsis CDK and CAK Orthologs

Our analysis suggested that there is no identifiable *CAK1/Csk1* ortholog in metazoa. If the CAK1 clade is ancient, that is, if it arose very early in eukaryotic evolution, it may have been subsequently lost in certain modern eukaryotic lineages such as metazoa. If this were the case

Table 8.1. Extended CDK family members used in this study

Genes	Species	Accession number
CDC28	*S. cerevisiae*	Z36029
KIN28	*S. cerevisiae*	X04423
PHO85	*S. cerevisiae*	Y00867
SGV1	*S. cerevisiae*	D90317
SRB10	*S. cerevisiae*	U20222
CTK1	*S. cerevisiae*	M69024
CAK1	*S. cerevisiae*	U60192
cdc2	*S. pombe*	AB004534
Mcs6	*S. pombe*	L47353
Csk1	*S. pombe*	S59896
SPSPCC16C4.11	*S. pombe*	AL031535
SPSPBC18H10.15	*S. pombe*	AL022304
SPSPAC2F3.15	*S. pombe*	Z99165
SPSPAC23H4.17C	*S. pombe*	Z98977
PI014	*S. pombe*	AB004534
T22H22.5	*A. thaliana*	AC005388
T12H1.1	*A. thaliana*	AC009177
K16E14.2	*A. thaliana*	AB026637
F21B7.1	*A. thaliana*	AC002560
AT4g22940	*A. thaliana*	AL161558
F8L10.9	*A. thaliana*	AC022520
F26A9.10	*A. thaliana*	AC016163
AT4g10010	*A. thaliana*	AL161516
F14J9.26	*A. thaliana*	AC003970
F6A14.22	*A. thaliana*	AC011809
F1M20.1	*A. thaliana*	AC011765
AAF21469.1	*A. thaliana*	U83118
T4P13.23	*A. thaliana*	AC008261
MXK3.19	*A. thaliana*	AB019236
K9H21.7	*A. thaliana*	AB023035
F12B7.13	*A. thaliana*	AC011020
F25P22.11	*A. thaliana*	AC012679
T10F20.5	*A. thaliana*	AC034107
CAK1At/AT4g28980	*A. thaliana*	AL161574
MBK5.8	*A. thaliana*	AB005234
CDC2a	*A. thaliana*	X57839
CDC2b	*A. thaliana*	D10851
At2g38620	*A. thaliana*	AC005499
F9H16.8	*A. thaliana*	AC007369
CaCAK1/CAC05182.1	*C. albicans*	AX005954
AC052571	*G. lamblia*	AC052571

continued on next page

Table 8.1. Cont.

Outgroup Taxa[a]		
KSS1	*S. cerevisiae*	Z72825
HOG1	*S. cerevisiae*	L06279
FUS3	*S. cerevisiae*	Z35777
SLT2	*S. cerevisiae*	X59262
MCK1	*S. cerevisiae*	M55984
MDS1	*S. cerevisiae*	U03280
YAK1	*S. cerevisiae*	Z49417

[a] Outgroup taxa belong to the *S. cerevisiae* ERK family (*KSS1, FUS3, SLT2, HOG1*), GSK3 family (*MCK1, MDS1*), and CLK family (*YAK1*) of kinases.[26]

then we would expect to see it in other eukaryotic kingdoms. Vascular plants are predicted to have diverged from the major eukaryotic lineage shortly before fungi diverged.[51] We therefore performed a phylogenetic analysis on the 98% finished *Arabidopsis thaliana* genome to search for a *CAK1/Csk1* ortholog.

We identified 24 *Arabidopsis* CDK family members by BLAST searches. Phylogenetic analysis groups them into five major clades, four of which contain yeast CDK family members (Fig. 8.3).

Arabidopsis cdc2a and *cdc2b* genes belong to the CDC28 clade. *cdc2a* is involved in cell cycle regulation. When a dominant cdc2a mutant was expressed in *Arabidopsis*, cell division was reduced.[52,53] *cdc2b* has been shown to be involved in cell elongation rather than cell division in hypocotyl growth in *Arabidopsis*.[54]

Arabidopsis MBK5.8 belongs to the SRB10 clade. Although nothing is known of MBK5.8's function, its yeast ortholog Srb10 and human ortholog CDK8 both function as CTD kinases that negatively regulate transcription.[55]

Arabidopsis F12B7.13 and *K9H21.7* both belong to the BC18H10.15 clade, which contains one fission yeast CDK family member *BC18H10.15*, as well as two *C. elegans*, two *Drosophila*, and two human CDK family members. Interestingly, this clade does not contain a budding yeast ortholog, suggesting it was lost (Fig. 8.3). The functions of the clade members are largely unknown, with the exceptions that the human orthologs *PISSLRE* and *PITSLRE* have been implicated in apoptosis and the G2/M cell cycle transition, respectively.[56-58]

One cluster of *Arabidopsis* CDK family members has no obvious ortholog in yeast. It branches near the SGV1 and CTK1 clades and contains 14 CDK family members: *T22H22.5, T12H1.1, K16E14.2, F21B7.1, AT4G22940, F8L10.9, F26A9.10, AT4G10010, F14J9.26, F6A14.22, F1M20.1, AAF21469.1, T4P13.23*, and *MXK3.19*. The biological functions of these genes have not yet been defined.

The KIN28 clade contains *Arabidopsis F25P22.11* and *T10F20.5* genes. F25P22.11 and T10F20.5 share more than 85% DNA sequence identity with each other, suggesting that they are derived from a recent gene duplication within plants.

Another *Arabidopsis* gene *cak1At* is located at the base of the KIN28 clade, although with insignificant bootstrap support (43% and 84% for NJ and ML analyses, respectively). A JTT distance matrix comparing Cak1At to the budding yeast CDK-family members indicates that Cak1At is most similar to Kin28p and is least similar to Cak1p (Table 8.2). Umeda et al showed that *cak1At* can complement both a budding yeast *cak1* mutant and a fission yeast

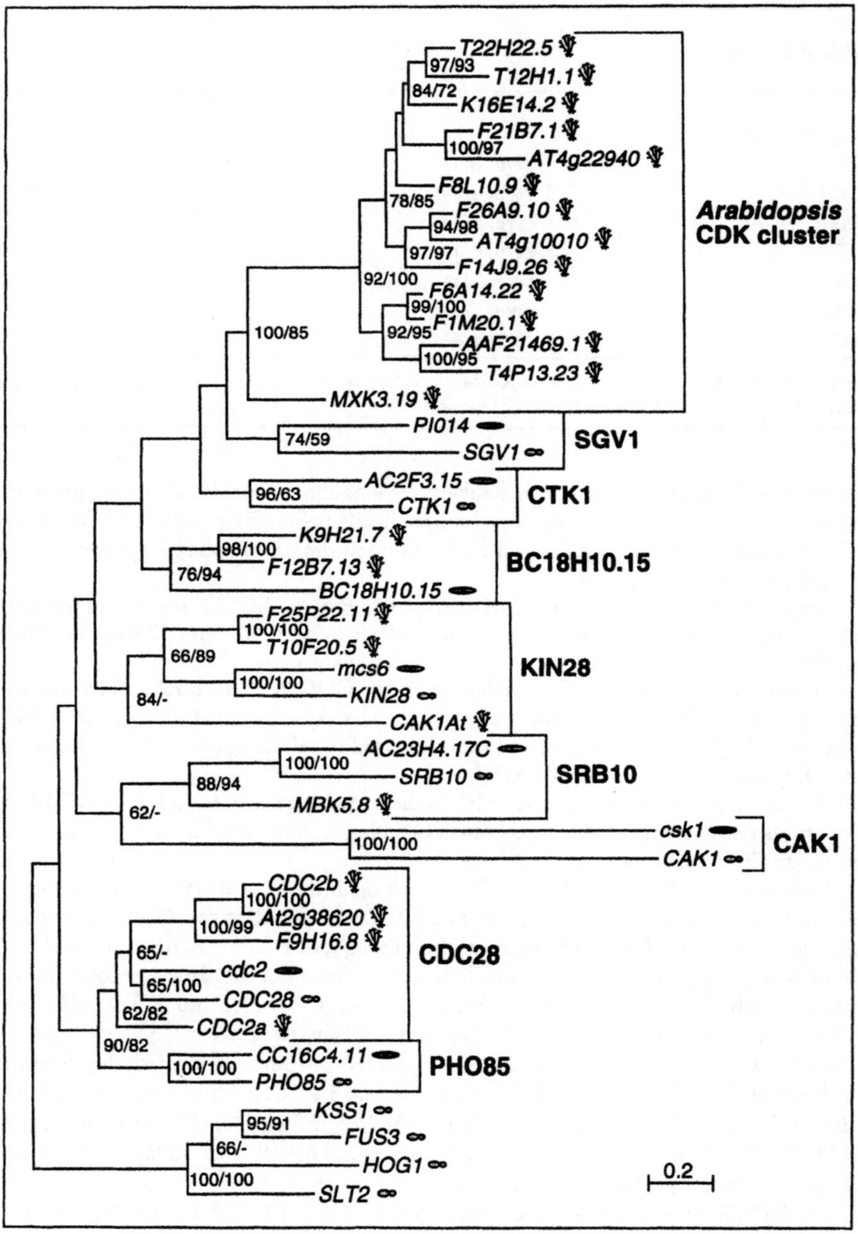

Fig. 8.3. Neighbor-Joining (NJ) and Maximum Likelihood (ML) phylogenies of *S. cerevisiae*, *S. pombe*, and *Arabidopsis thaliana* CDK family members. Branch lengths are proportional to the estimated number of amino acid substitutions; scale bar indicates amino acid substitutions per site. Bootstrap scores were obtained from 1000 replicates for the NJ analysis and 10,000 replicates for the ML analysis; scores above 50% are given at branch nodes (ML on the left and NJ on the right, separated by slash marks). Species are denoted by cartoon. Ancestral clades are denoted by brackets on the right. For a description of the methods see ref. 25. The sequence alignment used for this analysis is available upon request from the authors.

Table 8.2. Distance Matrix of Candida, Arabidopsis, and Giardia CAK orthologs vs. yeast CDKs

	Cdc28	Pho85	Kin28	Srb10	Ctk1	Sgv1	Cak1	Csk1
CaCAK1	2.17	2.10	2.16	**2.25**	2.16	2.12	***1.77***	1.73
Cak1At	1.23	1.39	***1.15***	1.40	1.42	1.41	**2.39**	2.18
T10F20 At	0.90	1.08	***0.80***	1.24	1.19	1.07	**1.94**	2.03
F25P22 At	0.88	1.05	***0.83***	1.22	1.18	1.05	**1.85**	2.02
AC052571 Gl	1.75	1.75	***1.37***	1.77	1.73	1.58	**2.40**	2.41

The scores represent the number of amino acid substitutions per site. Lower scores indicate taxa that are more similar, while higher scores indicate taxa that are more dissimilar. The scores of budding yeast taxa that are most similar to *C. albicans* CaCAK1, *A. thaliana* Cak1At, T10F20, and F25P22, and *G. lamblia* AC052571 are bold and italicized; those most dissimilar are bold and underlined. A column comparing fission yeast Csk1 is also given, but scores are not highlighted. The scores are JTT distance scores that were created with the protml program of the MOLPHY version 2.3.[64]

Mcs6 mutant, indicating that it has in vivo CAK activity.[59] They also showed that Cak1At has in vitro CAK activity towards Cdk2.[59] The authors classified Cak1At as a novel type of CAK, different from both Cak1p and CDK7, as Cak1At has a divergent sequence.[59] However, based on the JTT distance data and the 84% bootstrap support in our ML tree, we think that *cak1At* is more likely to be a divergent *KIN28/Mcs6/CDK7* ortholog.

There are no *Arabidopsis* CDK-family members that group with the CAK1 clade. Also, the only *Arabidopsis* orphan CDK family members, *cak1At* and members of the "*Arabidopsis* CDK cluster", are unlikely to be *CAK1/Csk1* orthologs, because both BLAST searches and the distance matrix showed that they are very dissimilar to CAK1 clade members. Therefore, with 98% of the genome sequenced, we have not observed a *CAK1/Csk1* ortholog in *Arabidopsis*.

Giardia lamblia CAK Orthologs

Giardia lamblia represents one of the most ancient eukaryotic lineages.[60] We searched the 4x sequenced *Giardia lamblia* genome by BLAST, PROFILE, and HMM searches[61] for a *CAK1/Csk1* ortholog and failed to find one. In contrast, a *KIN28* ortholog in *Giardia lamblia* (*AC052571*) was readily identifiable by BLAST search. The affiliation of AC052571 with the KIN28 clade is supported by both ML and NJ analyses with 95% and 81% bootstrap supports, respectively (data not shown), and is reflected in a JTT distance matrix (Table 8.2).

Three Hypotheses on CAK1 Evolution

Our analyses have failed to identify a recognizable CAK1 clade member in the plant *Arabidopsis*, or in the metazoa *C. elegans*, *Drosophila*, and humans. Assuming that a *CAK1/Csk1* ortholog is not in the remaining unsequenced 2% of the *Arabidopsis* genome, three possibilities for *CAK1/Csk1* evolution remain:

1) The CAK1 clade is ancient with members present in the major eukaryotic kingdoms, however, the clade members in distantly related phyla have diverged to the extent that identification by phylogeny is not possible.

2) The CAK1 clade is ancient and could be identified by phylogeny, however, its orthologs have been lost independently in both plants and animals.

3) The CAK1 clade originated in fungi and does not exist in any other eukaryotic kingdoms.

The first possibility suggests that *CAK1/Csk1* orthologs exist in metazoa and plants, but they are too divergent to be identified. One way to address this possibility is to assign all of the CDK-family members in a plant or metazoan species to known ancestral CDK clades and demonstrate either that *CAK1/Csk1* orthologs are present or are missing. We have been able to assign ancestry to the majority of metazoan CDK-family members with the exceptions of the orphan genes (Fig. 8.1).[25] The metazoan orphan genes are quite divergent from each other and do not have counterparts in *Arabidopsis*, suggesting that they are undergoing rapid evolution that is obscuring their relationships to ancestral clades. The sequencing of "missing-link" species may allow assignment of the ancestry of the metazoan orphan CDK family members, the members of the "Arabidopsis CDK cluster", as well as provide a definitive assignment for *cak1At*.

The second hypothesis predicts that the CAK1 clade is ancient but was lost independently in metazoa and plants. We have not found a *CAK1/Csk1* ortholog in the diplomonad *Giardia*, indicating that the CAK1 clade may not have existed in early eukaryotic lineages. However, it is still possible that the CAK1 clade is ancient but arose after the divergence of diplomonads from the main eukaryotic lineage. The test of this hypothesis will be determining whether there are *CAK1/Csk1* orthologs in taxa that diverged shortly before or after fungi did, such as the choanozoa.[62]

The third hypothesis, that the CAK1 clade originated in fungi, is consistent with our inability to detect *CAK1/Csk1* orthologs in other species. This hypothesis generates a prediction that can be tested. If the CAK1 clade arose in fungi, then it would have come from the duplication of a fungal CDK. Finding this gene duplication event would strongly indicate that the CAK1 clade arose in fungi. Ideally, if we identified a fungal *CAK1/Csk1* ortholog that shares significant sequence homology with one of the other fungal CDK genes, then this would indicate that the CAK1 clade derived from the ancestor of this CDK gene. A successful example of this type of analysis was the study of *bicoid*'s evolution.[63] *bicoid* is an essential homeobox gene and had been identified only in the closest relatives of the schizophoran fly *Drosophila*. Stauber et al cloned *bicoid* from a basal cyclorrhaohan fly, *Megaselia abdita*, and showed by phylogenetic analysis that the gene originated from a duplication of the *Megaselia* gene *zerknullt*, which is also conserved in vertebrates. Such an approach may be possible for the study of CAK1 evolution as well.

Summary

There are two categories of CAKs in eukaryotes encompassing the orthologous genes *CAK1/Csk1/CaCAK1* and *Mcs6/CDK7*. *Mcs6/CDK7* orthologs are highly conserved from *Giardia* to metazoa, indicating that this is an extremely ancient eukaryotic CDK clade, whose members are under functional constraints that limit their evolution. Members of this clade, including the budding yeast KIN28, function as central regulators of transcription.[32] The conservation of this clade may therefore be linked to its essential role in transcription rather than to its CAK activity. CAK activity for Mcs6/CDK7 orthologs has only been observed in fission yeast and in metazoa. It is an open question whether CAK activity for Mcs6/CDK7 orthologs is evolutionarily conserved beyond certain fungi and metazoa.

CAK1/Csk1 orthologs have not been identified in species other than yeast. The members of the clade appear to be evolving under relaxed evolutionary constraints as they are very divergent even within yeast. Currently it is not known whether the *CAK1/Csk1* genes are specific for fungi or whether their metazoan and plant orthologs have diverged to an extent that precludes their current identification.

Acknowledgements

We would like to thank R. P. Fisher for sharing information about the mouse CDK PNQALRE sequence; and G.L. Moulder and R. J. Barstead for providing *H01G02.2* knock out animals. E.T.K. is supported by grants from NIH (R01 GM55297) and the Human Frontiers Science Program Organization (RG-229).

References

1. Nasmyth KA, Reed SI. Isolation of genes by complementation in yeast: molecular cloning of a cell-cycle gene. Proc Natl Acad Sci USA 1980; 77(4):2119-2123.
2. Beach D, Durkacz B, Nurse P. Functionally homologous cell cycle control genes in budding and fission yeast. Nature 1982; 300(5894):706-709.
3. van den Heuvel S, Harlow E. Distinct roles for cyclin-dependent kinases in cell cycle control. Science 1993; 262:2050-2054.
4. King RW, Jackson PK, Kirschner MW. Mitosis in transition. Cell 1994; 79:563-571.
5. Sherr CJ. G1 phase progression: cycling on cue. Cell 1994; 79:551-555.
6. Stillman B. Cell Cycle Control of DNA replication. Science 1996; 274:1659-1664.
7. Morgan DO. Cyclin-dependent kinases: Engines, Clocks, and microprocessors. Ann Rev Cell Dev Biol 1997; 13:261-291.
8. Valay JG, Simon M, Dubois MF, Bensaude O, Facca C et al. The KIN28 gene is required both for RNA polymerase II mediated transcription and phosphorylation of the Rpb1p CTD. J Mol Biol 1995; 249(3):535-544.
9. Liao SM, Zhang J, Jeffery DA, Koleske AJ, Thompson CM et al. A kinase-cyclin pair in the RNA polymerase II holoenzyme. Nature 1995; 374(6518):193-196.
10. Lee J, Greenleaf, AL. CTD kinase large subunit is encoded by CTK1, a gene required for normal growth of Saccharomyces cerevisiae. Gene Expr 1991; 1(2):149-167.
11. Sterner DE, Lee JM, Hardin SE, Greenleaf AL. The yeast carboxyl-terminal repeat domain kinase CTDK-I is a divergent cyclin-cyclin-dependent kinase complex. Mol Cell Biol 1995; 15(10):5716-5724.
12. Prelich G, Winston F. Mutations that supress the deletion of an upstream activating sequence in yeast: involvement of a protein kinase and histone H3 in repressing transcription in vivo. Genetics 1993; 135(3):665-676.
13. Zhu Y, Pe'ery T, Peng J, Ramanathan Y, Marshall N, Marshall T et al. Transcription elongation factor P-TEFb is required for HIV-1 tat transactivation in vitro. Genes Dev 1997; 11(20):2622-2632.
14. Reines D, Conaway RC, Conaway JW. Mechanism and regulation of transcriptional elongation by RNA polymerase II. Curr Opin Cell Biol 1999; 11(3):342-346.
15. De Bondt HL, Rosenblatt J, Jancarik J, Jones HD, Morgan DO et al. Crystal structure of cyclin-dependent kinase 2. Nature 1993; 363(6430):595-602.
16. Russo AA, Jeffrey PD, Pavletich NP. Structural basis of cyclin-dependent kinase activation by phosphorylation. Nat Struct Biol 1996; 3(8):696-700.
17. Kaldis P, Sutton A, Solomon MJ. The CDK-activating kinase (CAK) from budding yeast. Cell 1996; 86:553-564.
18. Thuret JY, Valay JG, Faye G, Mann C. Civ1 (CAK in vivo), a novel CDK-activating kinase. Cell 1996; 86(4):565-576.
19. Hermand D, Pihlak A, Westerling T, Damagnez V, Vandenhaute J et al. Fission yeast Csk1 is a CAK-activating kinase (CAKAK). EMBO J 1998; 17(24):7230-7238.
20. Larochelle S, Pandur J, Fisher RP, Salz HK, Suter B. Cdk7 is essential for mitosis and for in vivo CDK-activating kinase activity. Genes Dev 1998; 12:370-381.
21. Kato JY, Matsuoka M, Strom DK, Sherr CJ. Regulation of cyclin D-dependent kinase 4 (cdk4) by cdk4-activating kinase. Mol Cell Biol 1994; 14(4):2713-2721.
22. Espinoza FH, Farrell A, Nourse JL, Chamberlin HM, Gileadi O et al. Cak1 is required for Kin28 phosphorylation and activation in vivo. Mol Cell Biol 1998; 18(11):6365-6373.
23. Kimmelman J, Kaldis P, Hengartner CJ, Laff GM, Koh SS et al. Activating phosphorylation of the Kin28p subunit of yeast TFIIH by Cak1p. Mol Cell Biol 1999; 19(7):4774-4787.

24. Nagahara H, Ezhevsky SA, Vocero-Akbani AM, Kaldis P, Solomon MJ et al. Transforming growth factor beta targeted inactivation of cyclin E:cyclin-dependent kinase 2 (Cdk2) complexes by inhibition of Cdk2 activating kinase activity. Proc Natl Acad Sci USA 1999; 96(26):14961-14966.
25. Liu J, Kipreos ET. Evolution of cyclin-dependent kinases (CDKs) and CDK-activating kinases (CAKs): Differential conservation of CAKs in yeast and metazoa. Mol Biol Evol 2000; 17(7):1061-1074.
26. Hanks SK, Hunter T. The eukaryotic protein kinase superfamily: kinase (catalytic) domain structure and classification. FASEB J 1995; 9:576-596.
27. Poon RYC, Yamashita K, Adamczewski JP, Hunt T, Shuttleworth J. The cdc2-related protein $p40^{MO15}$ is the catalytic subunit of a protein kinase that can activate $p33^{cdk2}$ and $p34^{cdc2}$. EMBO J 1993; 12:3123-3132.
28. Solomon MJ, Harper JW, Shuttleworth J. CAK, the $p34^{cdc2}$ activating kinase, contains a protein identical or closely related to $p40^{MO15}$. EMBO J 1993; 12:3133-3142.
29. Buck V, Russell P, Millar JB. Identification of a CDK-activating kinase in fission yeast. EMBO J 1995; 14(24):6173-6183.
30. Damagnez V, Makela TP, Cottarel G. Schizosaccharomyces pombe Mop1-Mcs2 is related to mammalian CAK. EMBO J 1995; 14(24):6164-6172.
31. Fisher RP, Jin P, Chamberlin HM, Morgan DO. Alternative mechanisms of CAK assembly require an assembly factor or an activating kinase. Cell 1995; 83(1):47-57.
32. Orphanides G, Lagrange T, Reinberg D. The general transcription factors of RNA polymerase II. Genes Dev 1996; 10(21):2657-2683.
33. Holstege FC, Jennings EG, Wyrick JJ, Lee TI, Hengartner CJ, Green MR et al. Dissecting the regulatory circuitry of a eukaryotic genome. Cell 1998; 95(5):717-728.
34. Feaver WJ, Svejstrup JQ, Henry NL, Kornberg RD. Relationship of CDK-activating kinase and RNA polymerase II CTD kinase TFIIH/TFIIK. Cell 1994; 79(6):1103-1109.
35. Cismowski MJ, Laff GM, Solomon MJ, Reed SI. KIN28 encodes a C-terminal domain kinase that controls mRNA transcription in Saccharomyces cerevisiae but lacks cyclin-dependent kinase-activating kinase (CAK) activity. Mol Cell Biol 1995; 15(6):2983-2992.
36. Lee KM, Saiz JE, Barton WA, Fisher RP. Cdc2 activation in fission yeast depends on Mcs6 and Csk1, two partially redundant CDK-activating kinases (CAKs). Curr Biol 1999; 9(8):441-444.
37. Felsenstein J. Cases in which parsimony and compatibility methods will be positively misleading. Syst Zool 1978; 24:401-410.
38. Lyons-Weiler J, Hoelzer GA. Escaping from the Felsenstein zone by detecting long branches in phylogenetic data. Mol Phylogenet Evol 1997; 8(3):375-384.
39. Saitou N, Imanishi T. Relative efficiencies of the Fitch-Margoliash, maximum parsimony, maximum likelihood, minimum evolution, and neighbor-joining methods of phylogenetic tree construction in obtaining the correct tree. Mol Biol Evol 1989; 6(5):514-525.
40. Felsenstein J. Evolutionary trees from DNA sequences: a maximum likelihood approach. J Mol Evol 1981; 17(6):368-376.
41. Saitou N, Nei M. The neighbor-joining method: a new method for reconstructing phylogenetic trees. Mol Biol Evol 1987; 4(4):406-425.
42. Yang Z. Among site rate variation and its impact on phylogenetic analyses. Trends Ecol Evol 1996; 11:367-372.
43. Brinkmann H, Philippe H. Archaea sister group of Bacteria? Indications from tree reconstruction artifacts in ancient phylogenies. Mol Biol Evol 1999; 16(6):817-825.
44. Lyons-Weiler J, Hoelzer GA, Tausch RJ. Relative apparent synapomorphy analysis (RASA). I: The statistical measurement of phylogenetic signal. Mol Biol Evol 1996; 13(6):749-457.
45. Altschul SF, Madden TL, Schaffer AA, Zhang JH, Zhang Z et al. Gapped BLAST and PSI-BLAST: A new generation of protein database search programs. Nucleic Acids Res 1997; 25:3389-3402.
46. Consortium TCeS. Genome sequence of the nematode *C. elegans*: A platform for investigating biology. Science 1998; 282:2012-2018.

47. Adams MD, Celniker SE, Holt RA, Evans CA, Gocayne JD et al. The genome sequence of *Drosophila melanogaster*. Science 2000; 287(5461):2185-2195.
48. Gribskov M, Luthy R, Eisenberg D. Profile analysis. Methods Enzymol. 1990; 183:146-159.
49. Johnson LN, Noble ME, Owen DJ. Active and inactive protein kinases: Structural basis for regulation. Cell 1996; 85(2):149-158.
50. Martinez AM, Afshar M, Martin F, Cavadore JC, Labbe JC et al. Dual phosphorylation of the T-loop in cdk7: its role in controlling cyclin H binding and CAK activity. EMBO J 1997; 16(2):343-354.
51. Schlegel M. Molecular phylogeny of eukaryotes. Trends Ecol Evol 1994; 9(9):330-335.
52. Hemerly AS, Ferreira PC, Van Montagu M, Engler G, Inze D. Cell division events are essential for embryo patterning and morphogenesis: Studies on dominant-negative cdc2aAt mutants of *Arabidopsis*. Plant J 2000; 23(1):123-130.
53. Hemerly A, Engler JdA, Bergounioux C, Van Montagu M, Engler G et al. Dominant negative mutants of the Cdc2 kinase uncouple cell division from iterative plant development. EMBO J 1995; 14(16):3925-3936.
54. Yoshizumi T, Nagata N, Shimada H, Matsui M. An *Arabidopsis* cell cycle-dependent kinase-related gene, CDC2b, plays a role in regulating seedling growth in darkness. Plant Cell 1999; 11(10):1883-1896.
55. Hampsey M, Reinberg D. RNA polymerase II as a control panel for multiple co-activator complexes. Curr Opin Genet Dev 1999; 9(2):132-139.
56. Lahti JM, Xiang J, Heath LS, Campana D, Kidd VJ. PITSLRE protein kinase activity is associated with apoptosis. Mol Cell Biol 1995; 15(1):1-11.
57. Ariza ME, Broome-Powell M, Lahti JM, Kidd VJ, Nelson MA. Fas-induced apoptosis in human malignant melanoma cell lines is associated with the activation of the $p34^{cdc2}$-related PITSLRE protein kinases. J Biol Chem 1999; 274(40):28505-28513.
58. Li S, MacLachlan TK, De Luca A, Claudio PP, Condorelli G et al. The cdc-2-related kinase, PISSLRE, is essential for cell growth and acts in G2 phase of the cell cycle. Cancer Res 1995; 55(18):3992-3995.
59. Umeda M, Bhalerao RP, Schell J, Uchimiya H, Koncz C. A distinct cyclin-dependent kinase-activating kinase of Arabidopsis thaliana. Proc Natl Acad Sci USA 1998; 95(9):5021-5026.
60. Smith MW, Aley SB, Sogin M, Gillin FD, Evans GA. Sequence survey of the *Giardia lamblia* genome. Mol Biochem Parasitol 1998; 95(2):267-280.
61. Eddy SR. Hidden Markov models. Curr Opin Struct Biol 1996; 6(3):361-365.
62. Embley TM, Hirt RP. Early branching eukaryotes? Curr Opin Genet Dev 1998; 8(6):624-629.
63. Stauber M, Jackle H, Schmidt-Ott U. The anterior determinant bicoid of Drosophila is a derived Hox class 3 gene. Proc Natl Acad Sci USA 1999; 96(7):3786-3789.
64. Adachi J, Hasegawa M. MOLPHY version 2.3: Programs for molecular phylogenetics based on maximum likelihood. Computer Science Monographs, No. 28. Institute of Statistical Mathematics, Tokyo 1996.

Index

GPSR Compliance

The European Union's (EU) General Product Safety Regulation (GPSR) is a set of rules that requires consumer products to be safe and our obligations to ensure this.

If you have any concerns about our products, you can contact us on ProductSafety@springernature.com

In case Publisher is established outside the EU, the EU authorized representative is:

Springer Nature Customer Service Center GmbH
Europaplatz 3
69115 Heidelberg, Germany

Batch number: 10682215

Printed by Printforce, the Netherlands